ROLLING SCULPTURE

AUBURN • CORD • DUESENBERG • STUTZ • FORD • LINCOLN • MERCURY • BUICK • STUDEBAKER • PACKARD • TASCO • WILLS ST. CLAIRE

ROLLING SCULPTURE

A DESIGNER AND HIS WORK

by GORDON M. BUEHRIG with WILLIAM S. JACKSON

ts
Haessner Publishing, Inc.
Newfoundland, New Jersey 07435

Library of Congress Catalog Card No: 75-10918
ISBN 0-87799-045-X

Designed by John M. Peckham

Haessner Publishing, Inc.
Newfoundland, New Jersey 07435

Printed in the United States of America

Dedication

This book is dedicated to the three women in my life:
My first wife, Betty Whitten Buehrig
Our daughter, Barbara Buehrig Orlando
My second wife, Kathryn Benzin Buehrig

Foreword

by
Richard A. Teague

In the automotive styling world, it is easy to generate differences of opinion. What is one designer's carriage is another's after-midnight pumpkin. Universal judgements of excellence are rare—and rarer still are those individuals whose genius inspires them.

Which is why it is so welcome that Gordon Buehrig, with the able collaboration of Bill Jackson, has taken the time to chronicle his remarkable career in automotive design.

In fact, it is more than a remarkable career—it is legendary. I have known Gordon for a quarter of a century, and known his work for more years than that. He is a founding father of automotive design, and has left an indelible mark upon it. He brought to the profession the insight, humaneness and creativity which makes legends. His ideas were unusual, fresh and stimulating—and what is most important, he brought them into being. This is a challenge which inevitably defines greatness. Many designers have ideas that merit acclaim. It takes something more to make them work.

The mark of the really exceptional car designer is the degree to which his creations are coveted and revered long after they were built. Many of Gordon Buehrig's cars are in this class—true collector's items. They were considered classic cars when introduced, and the feeling about them, the sense of distinction and value, has increased with the passage of time.

What discussion of classic cars could be held without including the Cord 810? For sheer boldness, dramatic appeal and innovation, it is now and always will be considered a piece de resistance by those for whom the automobile has a special meaning.

These remembrances by a towering figure will certainly be of interest to professionals, students and collectors alike. Gordon Buehrig can describe the special meaning, and how it is applied in a pragmatic business world, as few people could—because he has felt it and lived it as few people have.

Richard A. Teague
Vice President, Styling
American Motors Corporation
Detroit, Michigan
May, 1975

Contents

Preface

I was born in the small town of Mason City, Illinois on June 18, 1904, the son of Fred and Louise Buehrig; and with brother Donald attended public schools through high school.

This documentation of my work starts after graduation from the Mason City high school.

About three-quarters of the Twentieth Century is now history and many years from now this century will have a name. It's anyone's guess what it will be, but to me it is the **Automobile Age.**

Journalists and historians have written and will continue to write many volumes about automobiles. Why then am I writing this book? Because I believe I have a story to tell, based on over half a century of association with the auto industry. Examples of my work as a designer are described in great detail in the following pages.

Had you been a Duesenberg customer many years ago, I would have done my best to design a body for your car that was exactly what you wanted. In writing an automobile book for your library I have tried just as hard to please you.

Gordon M. Buehrig

Introduction

To the lover of the classic automobile, Stutz, Auburn, Cord and Duesenberg are remembered as among the finest. Their engineering and beauty are representative of the best in the classic automobile.

What a feeling of accomplishment must come to a man who can state he was chief of body design for all of them. Add to that a second design career in the modern era of the automobile which was only recently concluded with his retirement from the Ford Motor Company.

There is only one man who can claim such a distinguished record of automotive design in the United States . . . a record of creation which has given us the fine body designs of the Duesenberg Model J, the 1935 Auburn Boattail speedster and the Models 810 and 812 Cord, the latter of which the Museum of Modern Art of New York City in 1951 called one of the eight finest automobile designs ever.

Gordon Miller Buehrig, whose individual automotive design creativity must be ranked with the greatest, is that man.

That Gordon Buehrig was able to individually create one outstanding design of the Classic Era would be significant in itself. That his Duesenbergs, Auburn boattail speedster and Cords are **all** considered among the most significant and esthetically appealing of that era makes his career most unusual.

Like many men of truly great accomplishment, Gordon Buehrig is not terribly impressed with himself. He has almost always loved his work. Throughout his career, integrity seems to have had more weight than financial security.

Judge for yourself. We give you here the story of one of our country's greatest automotive designers and the story of the great cars he designed.

William S. Jackson

ROLLING SCULPTURE

A beginning 1

I graduated from high school in 1922 and entered Bradley Polytechnic in Peoria, Illinois, where I took a liberal arts course with some art. One day I was expelled from the chemistry class because, according to the instructor, my notebook was not well done and there were sketches of automobiles on all the pages. He was incensed at the thought I wasted my time drawing automobiles.

When summer vacation came, I went to Chicago and got a job driving a Yellow Cab. This at least satisfied my love of driving and there were some elegant automobiles to be seen in the Windy City.

Chicago had two custom body firms at that time. One was named Graff and the other was the C.P. Kimball Company. One afternoon in November 1923, I stopped at Kimball and asked to see the designer. Clarence Wexelberg was the designer, body engineer and layout draftsman for Kimball. During my interview he graciously told me how to plan my education and my career if I really wanted to design automobiles. In January 1924 I went back to Bradley and enrolled in courses in drafting, wood shop, metal shop and art.

The "Hoopus," Gordon Buehrig's first automotive design work. He, and brother Don, built this interesting roadster from a 1904 Orient Buckboard.

In the summer of 1923, Gordon Buehrig went to Chicago and got a job driving a Yellow Cab. He was wearing this uniform when he had his first interview with Clarance Wexelberg at the C.P. Kimball Company there.

Gotfredson Body

When the school term ended in May 1924, I left college and went to Detroit, following the plan Clarence Wexelberg had outlined for me. I had written and obtained interviews with two men in the automobile body business. The first, Eugene Botzenhardt was body engineer of the Fisher Body Company. The second was Walter Jones, chief engineer of the C.R. Wilson Body Company.

At the time, Jones was in the process of changing jobs, having been hired as chief engineer of the Gotfredson Body Company in Wayne, Michigan. He hired me to be an apprentice in the sample body shop. I started work there November 8, 1924. I found that my shop courses at Bradley were good training for my first job in the automobile industry.

Our production commitments at Gotfredson included touring cars for Jewett, a sedan model for Peerless and the roadster and sedan models for Wills Sainte Claire. Virtually all of these bodies were of composite construction.

I was not in the wood shop long enough to acquire much skill as a craftsman but I did learn to understand that part of the business. After only six months of apprenticeship, I was moved from the woodshop to engineering to become a body draftsman. Bill Jones, Walter's brother, was chief draftsman and I started my apprenticeship in engineering under his capable leadership. Drafting and engineering were even more fascinating to me than work in the sample shop because I was one more step closer to design.

The drafting room at Gotfredson Body Company, Wayne, Michigan, where Gordon Buehrig officially entered the world of automobile design on November 8, 1924. Note the blackboard drawing against the far wall - a Peerless Sedan. This photograph was taken in Spring 1926.

The 1925 Wills Sainte Claire Fixed Windshield Roadster was one of the bodies being built at Gotfredson Body Company, Wayne, Michigan, when Gordon Buehrig joined the firm in November 1924.

Dietrich Incorporated

On January 14, 1926, I changed companies. I got a job as a draftsman in the engineering department of Dietrich Incorporated, a custom body firm in Detroit. Dietrich had been formed in 1925 when designer Ray Dietrich had left LeBaron to go on his own. Financial backing for this move had reportedly come through the Murray Corporation, another body builder in Detroit, who was closely associated with the Lincoln Division of the Ford Motor Company.

My life in the automobile industry became more exciting with every move. Here we were building bodies for such companies as Pierce-Arrow, Franklin, Packard, Lincoln and occasionally a custom body for a foreign chassis either on an individual basis on request, or on a contract basis. The three largest builders of prestige automobiles at the time were Packard, Lincoln and Cadillac.

Dietrich bodies, like those of other custom builders, had certain construction characteristics that were unique to that particular builder. For instance, on Dietrich bodies the windshield pillar was usually a bronze casting that was extremely narrow. The width of the pillar was less than the distance between your eyes and therefore was, theoretically, not a blind spot. It was a safety feature invented by Ray Dietrich.

Packard

I started working for Packard on January 16, 1927. It was a great place to work in those days. They built fine automobiles. Their employee turnover was extremely low and the people who worked there were proud of their company.

At this time I was not a very experienced body layout man and my work on the full-sized boards was largely confined to making drawings of the body panels.

While working for Packard, I met a gentleman to whom I will always be indebted for the knowledge gained through his friendship. Frederick J. Hooven gave me a book by LeCorbusier entitled "Toward a New Architecture." This great book helped me formulate much of the philosophy of design which I used later in designing automobiles.

Another acquaintance I made the same year had an immediate effect on my future. Joe Pizzo had just been hired by Harley Earle, who was heading up a new central design department for General Motors. Through Joe's influence, I had an interview with Mr. Earle the following week and was employed as an AUTOMOBILE DESIGNER!

General Motors

I started work for the Art and Colour Department of General Motors. I was the fourth designer hired to work in the new department. Prior to the formation of Art and Colour, each division of General Motors had its own design department. There was little if any interchangeability of body parts among designs except for body hardware.

The significance of this move by General Motors was far greater than realized at the time. Fisher Body was changing over from composite body construction to all steel construction and the resulting tooling costs were going up at a rapid rate. By having a central design department, Harley Earle and the various divisions of General Motors were able to put together a plan whereby Chevrolet, Pontiac, Oldsmobile, Buick and Cadillac were able to use three basic body shells. This was a brilliant plan and amortization of the tooling costs was apportioned to the various divisions based on the number of cars they produced. This merchandising scheme was probably one of the main factors of General Motors' success during and after the depression of the Thirties.

The automobile being designed when I started with GM was the 1929 Buick. The full-sized clay model of the car was almost finished and it was my first experience seeing an automobile sculptured in clay. The technique was so new that a special type of clay imported from Germany was used. Later, a firm in New Jersey was established to provide a domestic source.

A 1929 Buick 5-passenger four-door sedan on a 121-inch wheelbase. The 1929 Buick was the first car Gordon Buehrig worked on when he joined General Motors design department.

The only part of the Buick I designed was the instrument panel, but it was a great satisfaction to me to have been on the design team of that car.

The 1929 Buick is the one dubbed "The Pregnant Buick." It was not one of the greatest designs ever, but there is a little-known aspect of its styling that should be put in the record.

The original design model which Harley Earle showed the executive board was an outstanding example of automotive art. I will describe it.

Buick, in their plans for 1929, included two basic series with two sizes of engines. The larger series also came in two wheelbases. The higher production models were the small engine series and the shorter wheelbase version of the larger engine series. However, the model which had the best proportions from an aesthetic standpoint was a four-door, blind quarter sedan on the long wheelbase chassis.

The rear of the body was only slightly aft of the rear axle and a fabric covered trunk was mounted to the body. The detail design of the greenhouse was the outstanding feature. The door pillars, above the belt, were very thin in cross section and could not be built by Fisher's standard door construction. The roof was covered with fabric. It was the same material used on convertibles, but in this case it was put over the roof for an aesthetic effect similar to the vinyl over steel roofs now popular on some cars.

A portion of the roof extended down to the belt between the front and rear doors. Had the car been produced exactly like the design model, it would have been a styling sensation. When the model was turned over to Fisher Body, they made numerous changes, including redesign by their engineers of the greenhouse.

In the 1938 model year, Cadillac brought out a model called the "60 Special" with a greenhouse almost identical to the 1929 Buick, and today it is a collector's item. Unfortunately, I have no pictures of the 1929 Buick design model as it was submitted and the foregoing description is strictly from memory.

This view from the 1929 Buick catalog shows in context the sheer simplicity of the instrument panel, which was the only part of the car Gordon Buehrig claims as his own, as he was a member of the design team which did the entire car.

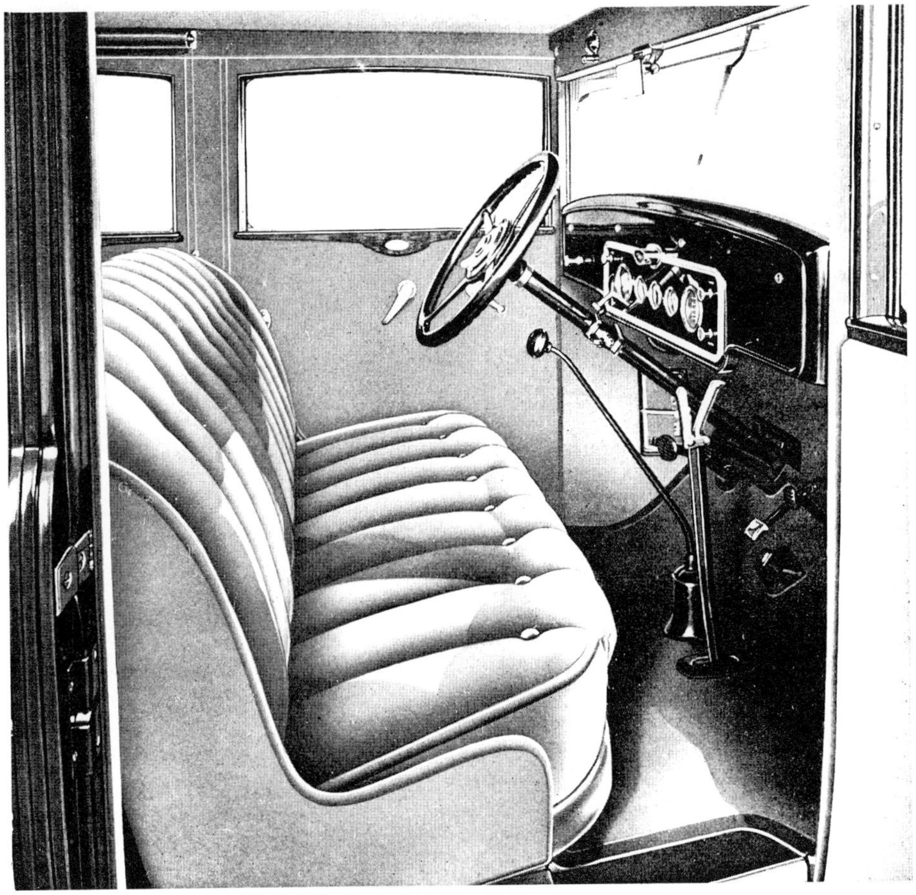

From the 1929 Buick catalog, a detail picture of Gordon Buehrig's instrument panel. A 90 mile-an-hour speedomoter was directly in front of the driver. A hydrostatic temperature gauge is on the right.

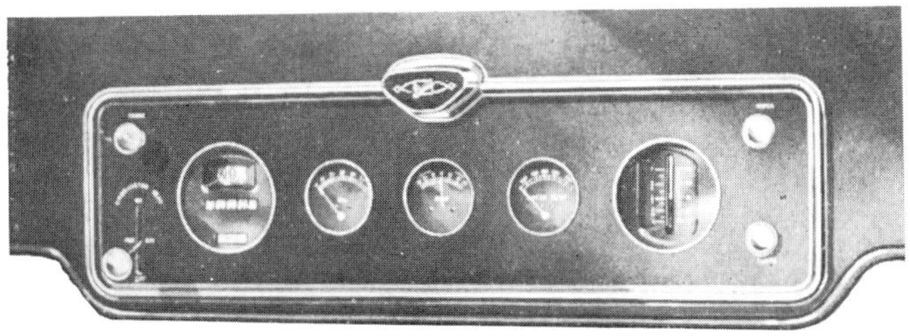

This was the first model designed by Harley Earle's new Art and Colour Department. Later he became a vice president of General Motors and had much more influence in the company. After that, few changes were ever made from an approved design.

Stutz | 2

I was a very naive young man. I was not smart enough to realize the fortunate position I was in at General Motors. I quit a few months after joining them to take a job as the body designer for the Stutz Motor Car Company of America in Indianapolis, Indiana. I left the largest and strongest automobile company to work for a small company which was in financial trouble and, as history proved, had only a short time to live.

When I arrived at Stutz in December of 1928, I found my position offered me relatively little opportunity for design work because the company did not have the financial capability to underwrite a new body program.

They had just introduced a new car called the Blackhawk to be a companion car to the regular Stutz. It was an attempt to grab a share of the medium-priced market and it reflected poor judgement on the part of management. Fred J. Moskovics had been president of Stutz, but he left shortly after I started and Colonel Edgar S. Gorrell, a World War I flyer and associate of Moskovics, was the new president.

The new series of bodies had been designed by Ralph Roberts and his design staff at the Briggs Manufacturing Company, the large body company in Detroit. The Stutz and the Blackhawk cars used the same bodies. The enclosed bodies on the Stutz were quite good looking and had fine proportions. The wheelbase was considerably longer than on the Blackhawk and the hood was also longer. The Stutz had a large straight eight-cylinder engine. The Blackhawk had a six. Both engines had single overhead camshafts. The Blackhawk was also available with a conventional Continental eight-cylinder engine, which was common equipment on a number of very ordinary cars. It clearly did not belong in a vehicle from the Stutz factory. On the Blackhawk, the new bodies looked too big for the car and the overall proportions were bad when compared with the Stutz. From a standpoint of performance, the Stutz was exciting and the Blackhawk was not.

During my days at Stutz I shared an apartment with Jack Beaty, a young man who worked in the accounting department at Stutz. Through him I learned about production costs. The retail price on the Stutz was something over $3,000 and the Blackhawk was about $1,000 less. However, the cost of producing the Stutz was only $200 more than the Blackhawk. The existance of the Blackhawk took away a great deal from the prestige of the name Stutz and hurt sales. In spite of the great difference in selling price, the Stutz sold reasonably well and there was virtually no market for the Blackhawk.

The Blackhawk was a fatal mistake which eventually put Stutz out of business. All businessmen sometimes make mistakes, but in the automobile business, just one like the Blackhawk is enough to kill a small company.

Stutz was ready to introduce a new engine called the DV32, which was a twin overhead camshaft type with four valves per cylinder. It was similar in layout to the Duesenberg Model J engine. Fred Duesenberg was the type of man who would help a competitor and I was told he sent Stutz complete prints of his engine to study. The new engine was a shot in the arm for the company and received good press reports but it was not enough to overcome the Blackhawk mistake.

My accomplishments in design at Stutz were quite limited. A Frenchman named Edouard Brisson, along with co-driver Robert Bloch, had taken second place in the 1928 LeMans 24 hour race in France, using a Stutz Blackhawk speedster. Their 68.78 mile per hour average brought them second only to a factory entered Bentley racing car. The Stutz had been entered and was owned by Charles Weymann, the famous French coachbuilder who was associated with Stutz in America.

Weymann ordered three special new cars for the next year. The chassis engineers laid out a beautiful sports car chassis by using a Blackhawk frame and installing the Stutz engine several inches aft of its normal location. The radiator was positioned about in line with the front axle. A Roots type supercharger was mounted in front of the radiator and was driven off the front end of the crankshaft. The straight-eight engine was bored to 3 3/8 inches, giving 322 cubic inches displacement and 115 horsepower at 3,600 rpm. When the supercharger was cut in, horsepower increased to 155 at 3,600 rpm. A new four-speed gearbox was also fitted.

LeMans 1929: Guy Bouriat and Baron Rothschild (George Philippe) in their Stutz #5.

Stutz #6 at speed on Le Circuit, LeMans, 1929. This was one of the Vaiture team cars.

It was my job to design the bodies and this was an exciting assignment. The international rules for racing sports cars that year called for a four seater open car and the minimum dimensions for the front seat were spelled out, but the rear seat layout was vague. Since only the driver was going to be in the car, I cheated very much on rear seat dimensions. Probably two children would have been comfortable in the rear.

The bodies were built by the Weymann American Body Company. They were constructed with a wooden frame which was only about one-third the weight of typical American construction. The joints were made by using metal plates screwed into the wood pieces in such a manner that there was a small gap between the wood parts. This completely eliminated noise and provided a certain amount of compliance to the bodies. They were only good on cars which had strong rigid frames.

The skin surface was made by first fastening a chicken wire cover over the framework. Then, a cover of padding and finally a skin surface of

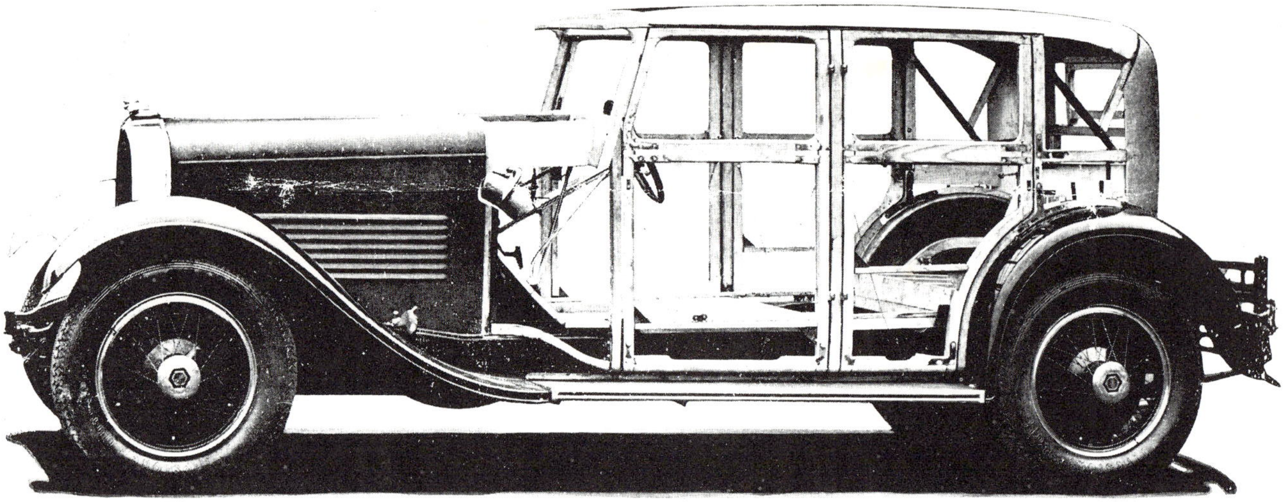

The photograph of an incomplete Weymann-bodied Stutz shows the unique body construction used. Note all wood pieces are held apart by steel brackets, giving flexible joints as in aircraft construction of the period.

imitation leather. The bodies were extremely light but were somewhat limited in shape because of their construction. I kept this in mind and designed the bodies to be compatable with the Weymann type of manufacture. The completed cars were very attractive.

However, 1929 did not prove to be a Stutz year at Le Mans. Three cars were entered to be driven by Edouard Brisson, another by Guy Bouriat and his co-driver "George Philippe" (who was actually the Baron Philippe de Rothschild) and the third for Englishman George Eyston. Both the Brisson and Eyston cars retired with ruptured fuel tanks and the Bouriat/Philippe Stutz finished fifth overall at an average speed of 64.8 miles per hour. The cars didn't do very well, but I don't think it was because of the bodies.

The 1928 Stutz and Blackhawk enclosed bodies were rather well designed and, as mentioned, on the Stutz chassis they had good proportions. They also had a roadster with a rumbleseat, a phaeton and a speedster which were poorly designed. The public and the dealers felt the same way. I was asked to improve the design with a minimum of expense.

At this time many of the most beautiful of the European sports cars featured cut down front doors, which were copied on the Stutz design. The proportions, while good on several of the European cars, did not look at all good on the Stutz. The windshield on the Stutz was the folding type and was located too far forward. The cockpit area was also too large. Getting into the small European roadsters was like putting on a suit just your size. When you sat in the Stutz, it was more like sitting in a bathtub.

I redesigned the cowl and front doors and eliminated the cut down door effect. I located the windshield closer to the driver and made the cowl deeper. All this had the effect of wrapping the car around the driver and front seat passenger a little more, giving them a more intimate relationship with the machine.

I designed a new instrument panel which was used. Stutz had a fancy engraved panel, covered with meaningless decoration and hard-to-read gauges. My design was a mechanical looking panel that seemed to belong on a Stutz. The Northern Engraving Company of La Crosse, Wisconsin, built the sample panel for us.

I enjoyed my days at Stutz. Had the company been financially sound, it would have been a good place for me. I was proud of the Stutz car and its reputation and I liked the people in the company. The Stutz chassis, at that time, was probably the neatest piece of machinery from the bodybuilder's viewpoint that was available in the United States. The

Gordon Buehrig did not like the cut down door effect on the 1928 Stutz and as shown here on the 1930 Blackhawk roadster. The design was changed on the 1930 Stutz.

In February 1929, Gordon Buehrig visited the Le Baron plant in Detroit to discuss design changes for the Stutz. While there, he drew some sketches of a sport convertible on the short wheelbase. He was unable to sell the idea of its production to management, but later—in 1930—a very similar car was introduced by Stutz, called the Super Bearcat. The one shown here is a little-changed 1933 version.

frame was low to the ground and rugged. Stutz used a worm-drive differential with the input shaft located several inches below the centerline of the rear axle. With this arrangement, it was possible to build lower and better proportioned vehicles than our competitors, most of whom were using straight-cut pinion gears in their differentials. The hypoid differential, which accomplishes about the same purpose as the worm-drive packagewise, was just coming into use. It proved more efficient.

For years Stutz engineers had shown imagination in design and courage to be different. The "Safety Stutz" which came out in 1926 was exciting and advanced. It had parallel strands of wire embedded in the windshield and as such was the first car to offer safety glass. They had heavy gauge steel running boards designed to give protection in a sideswipe accident. The low center of gravity of the chassis gave the car good stability and handling.

Duesenberg memories | 3

Somewhere I had learned Duesenberg was looking for a designer.

A few days after the 1929 Indianapolis 500 Race, I had an interview with Harold T. Ames, the vice-president of Duesenberg Incorporated. I was greatly impressed by him and very excited when he hired me to be their body designer. There I was, still only 25 years old, and the chief body designer for the most prestigeous automobile ever built in the United States!

Harold T. Ames was vice-president of Duesenberg Incorporated when he interviewed Gordon Buehrig for the position of chief designer in June 1929.

E.L. Cord, the mastermind who created the Cord Corporation and gave us some of the finest cars of the Classic Era.

For the first three years with Duesenberg, I was the only body designer at the factory. Herb Newport and Alex Tremulis were employed later, but at a period after I had designed the cars for which I here take credit.

Duesenberg purchased all of the bodies for the Model J series from custom body companies for the entire run from 1928 through 1937. The regular chassis sheet metal was common on most of the cars, although there were a few that had custom built fenders and other items.

The only cars I designed with different fenders were two boattailed speedsters. Actually, I admired the standard equipment so much I felt sure I could not improve on it.

The items to which I am referring were included with the chassis: fenders, running boards, radiator shell, hood, headlights, taillights, bumpers, horns, instrument panel, etc. These items were all designed by

The Model J Duesenberg assembly area is shown here in Fall 1928, just after production of the Model J had begun.

The Duesenberg Administrative Building, with showroom in front, as it looked at about the time Gordon Buehrig joined the company.

Harold Ames and Errett Cord. Neither of these men could sketch or make orthographic drawings, but they could visualize what they wanted and were able to get it by working with draftsmen and pattern makers.

Mr. Cord and Mr. Ames should also be given credit for the fact all of my designs which were built were approved by them. Ames and Cord both had intuitive sense for the value of design and knew what was good.

During my years as chief body designer at Duesenberg, I was responsible for about half of the designs produced. Here I will only cover in detail the Duesenberg bodies I designed.

There have been thousands of words written about Fred and August Duesenberg. My association with Fred was unique. After I had been with the company about a year, he invited me to live at his home, which I did for over three years. He seemed old to me at the time and in fact was about twice my age, yet he died at the age of 54, which now seems quite young.

Historically, Fred and August Duesenberg are presented as Fred, the designer and genius, and Augie as the builder. This is generally correct, but as I was to learn after Fred's death, Augie also had considerable design talent.

When E.L. Cord set up the company of Duesenberg Incorporated in 1926, he hired only Fred. Augie therefore was not involved in the design or development of the model J Duesenberg. During these years Augie

Fred Duesenberg here helps a young lady into a LeBaron phaeton.

had a small racecar shop across the street from the Duesenberg factory and here he built and campaigned his race cars. After Fred died as a result of an automobile accident on July 26, 1932, Ames hired Augie and he was with the company for the rest of its existence.

When the Model J Duesenberg was introduced in 1928 the economy of the country was booming. The stock market was bullish and the situation could not have been better for the introduction of a high-priced luxury car. It was Harold Ames' responsibility to procure the bodies for the Duesenberg and he took the logical course of having the best custom body builders submit designs. Two of these were outstanding. The convertible coupe by Murphy and the phaeton by LeBaron that were shown at the New York Auto Salon are still considered to be great automobile art. An excellent example of the LeBaron phaeton is housed in the Briggs Cunningham Collection in California and a number of Murphy convertible coupes have been restored by collectors.

A large part of the fine car market in 1928 was for chauffer-driven limousines and town cars. Ames selected these models from designs submitted by different custom body companies. His volume commitments were conservative and had the stock market crash not occurred in November 1929, these bodies would have been used in a short time.

In late 1931, Gordon Buehrig discussed the possibility of a new catalog for the Model J Duesenberg, which would include many of the bodies since the first catalog was issued in 1928. He, Carl Otto and Helen Briggs completed these illustrations for the new catalog in the five days from November 6 to 10, 1931. Two examples are shown. The one below is of a Buehrig design. On the opposite page, the Convertible Victoria was a Rollston design.

This car is suitable for owner or chauffeur driving. Designed on long wheelbase and with partition. Auxiliary, side facing seats. A large, roomy body; may be used either as a sedan or phaeton.

THE DUESENBERG CONVERTIBLE BERLINE — DESIGN NUMBER 47B502

4

An understanding of the American economy at the time and in particular as it affected the market for Duesenbergs is necessary to appreciate the remarkable job Harold Ames did in keeping the company alive for the next eight years.

Reflect for a moment on these figures. On September 3, 1929, the Dow Jones Industrials average was 381.17. On June 30, 1932, the figure was 43.

In 1929, before the crash, the country was loaded with "millionaires on paper." Stocks could be purchased with only a ten percent margin and in a steady "bullish" market it was easy to pyramid one's holdings into fantastic figures and become rich **IF** you sold your stock.

When the market crashed, most investors who gambled in the market on margin could not cover their bets and were wiped out.

When Mr. Ames made his original pre-crash commitments for bodies for the Model J Duesenbergs, he got from most of the custom builders bodies which were similar to his competition. Before the market crash Duesenberg competed pricewise with Rolls-Royce, Hispano-Suiza, Minerva, Mercedes Benz, Isotta Fraschini and Bentley. After the crash, American buyers were looking at cheaper cars, for even if you had retained your wealth, it was not considered good taste to flaunt it when thousands of Americans were standing in bread lines. Consequently, Duesenberg customers were also shopping around for Packards, Lincolns, Cadillacs, etc. Some wealthy people were even buying Model A Ford town cars!

The top folds compactly, making a very attractive open car. The individual front chairs are unusually comfortable, and there is exceptional room in the rear, due to the sunken floor.

THE DUESENBERG CONVERTIBLE VICTORIA DESIGN NUMBER 1051101

5

Now that Duesenberg had to compete with lower-priced American fine cars, it was very detracting to Duesenberg sales to have virtually the same bodies as were available on cars selling for less than half the price. The Duesenberg bodies from the original commitments were held in storage at the various body companies and Ames made a special effort to sell them.

The Duesenberg bodies I designed were built by these same body builders but they were not available on any other chassis. Several of these models were quite popular but they were ordered in very limited quantities while the earlier bodies were still in inventory.

For example, suppose you were a potential customer for a new limousine and you were invited by a Duesenberg salesman to visit the Auto Salon. He might show you a Willoughby limousine Duesenberg with a price tag of $12,000. While viewing the rest of the show you might see a Willoughby limousine Lincoln with substantially the same body, but priced at $5,000. On the other hand, if you saw a Duesenberg Beverly Berline by Murphy—one of my designs—priced at $14,000 and there was nothing else like it in the show, you might buy it.

The Model J Duesenberg chassis was produced in two wheelbase lengths. The short chassis was 142.5 inches and the long chassis had a wheelbase of 153.5 inches. Chassis sheet metal was the same on both except for the running boards and splash aprons, which were eleven inches longer on the 153.5 inch chassis.

There were also two special short chassis made by cutting the frame which had a wheelbase of 125 inches. These were made for two-passenger bobtailed speedsters. These were built during the period in 1933 when I was working for General Motors. They were special orders for movie stars Clark Gable and Gary Cooper. The body design for these cars was executed by J. Herbert Newport. During my design days at Duesenberg, I designed two short chassis special bodies which were not built.

I made most of my design sketches in India ink but some were wash drawings using Chinese ink. I made a few attempts in watercolor wash and tempra but I had an insufficient amount of talent and training to handle those mediums in a professional manner.

My favorite was India ink and in this medium I used a trick to lighten the work load. I carefully made a drawing of the short wheelbase chassis and the chassis sheet metal in side elevation in one sixteenth scale. I gave it a hint of perspective by moving the front and rear wheels in the foreground one inch forward (in scale) and the rear wheel one inch aft. The wheels in the background were moved closer together by the same amount. This still left the centerline of the car in true scale.

Then photostats of the drawing were made and on one of them the drawing was cut and extended to the dimensions for the long wheelbase chassis. A number of these prints were ordered in both wheelbase lengths and by using them all I had to do was design the bodies. On the finished sketch, the chassis was just a phototstat and the body was India ink. This was not attractive but a second photostat of it looked quite professional.

Automobile designers have been notorious for making sketches of cars that glamorize their designs by cheating on proportions and roof lines. Most five-passenger cars of this period located the rear seat over the rear axle. Because of the vertical travel of the differential when the car hit a bump in the road, the rear seat was usually higher above the ground than the front seat. This was compensated for by a roof line that ran uphill toward the rear and was not a pleasing line. Most designers would cheat this line on their sketches.

The long wheelbase Duesenberg chassis was planned for seven-passenger cars and the short chassis was for five-passenger bodies. I designed most of my five-passenger bodies on the long wheelbase chassis. By doing this the rear seat was forward of the rear axle and could be the same height as the front seat. This arrangement provided an optimum condition for a perfect roof line on the vehicle and there was no reason to cheat on the design sketch, which was one thing I wouldn't do.

Today's five-passenger cars usually have the rear seat forward of the rear axle and the arrangement has been achieved by using a V-8 engine instead of a straight eight and by moving it forward in the chassis to provide more body space.

An orthographic side elevation design drawing also does not give a true picture because it eliminates foreshortening due to perspective. It is possible to give just the right amount of perspective to acquire a true picture and I always tried to do this.

The next step after completing the design sketch was to obtain Harold Ames' approval. Following this, I made a one-eighth scale body draft of the design. The importance of this step cannot be overemphasized. It was in the empirical design lines of the scale body draft that the actual shape of the body was specified. A body designer without a background in body engineering and body drafting would be unable to do this.

If the scale body draft was not made and only copies of the design sketch were sent to three body builders for quotations the results would be as follows: If Derham received the contract, the resulting body would have contour characteristics of a Derham body. If Brunn did it, the body would have a Brunn character and if Murphy did it, the body would be distinctively a Murphy body.

By using the scale body draft method it didn't make any difference which builder got the contract because, as far as the body contour was concerned, it would be exactly the same: it would be a Buehrig design.

This point was clearly proven in two instances where the same designs were built by two builders. The first instance was my Beverly design. The first ten Beverly Duesenbergs were built by Murphy in Pasadena, California, and the second two by Rollston in New York City. Without looking at the builder's nameplate on the sill panel it is almost impossible to tell a Murphy Beverly from a Rollston. On one of the Rollston Beverly, the "V" between the doors was eliminated. It was built for Mrs. William Wrigley and the change was probably made at her request.

The second instance was the Torpedo Phaeton. The first one was built in Buffalo by Brunn and the other four by Weymann in Indianapolis. Weymann built bodies of conventional construction as well as their patented fabric-covered type. Here again, there are no Brunn or Weymann characteristics and the body shape on each is identical. They are one hundred percent Buehrig design. Two of the four Weymann Torpedo Phaetons carried the Weymann body plate, while two carried a Walker-LeGrande plate. All four were built by Weymann.

The method of developing a body surface is a complex form of geometry, well known to all full-sized body draftsmen, but it is too complex for this book. The important point in the process to be mentioned here is before the surface lines are developed, they must be generated from certain design lines. In the method I used, all these design lines were shown on my scale body draft.

On page 52 of this book, a description of body drafting and surface development is included which barely scratches the surface of the subject, but does give a very general idea of this complex process. In recent years, even this process has been changed with the use of computer technology.

4 Duesenberg . . . the Buehrig bodies

When I left Duesenberg, I didn't take any original design work with me. I did take copies of photographs of cars and reproductions of some of my design sketches and scale body drafts. In the past forty-odd years much of this material has either been lost, given away or destroyed. Consequently, complete design details for all of my work at Duesenberg cannot be shown. In some cases, there are only photographs and my design sketches. For a few there are reproductions of the small scale body drafts. It is as complete as possible.

Shreve Archer Judkins Coupe

The first Duesenberg Model J body I designed was strictly a custom car. It was for Shreve Archer, an industrialist from Minneapolis, Minnesota. Archer wanted us to build him a simple two-passenger coupe that would be as unostentatious as any body could be on a Duesenberg chassis.

The car was painted black and had black tires instead of white sidewalls. It had no ornamentation except a chrome drip molding above the doors which joined with a chrome bead outlining the rear quarter.

The body was built by the J.B. Judkins Company. The car had good proportions and was elegant in its simplicity. However, no additional ones were built to my knowledge, because most customers desired more unusual styling.

The Shreve Archer Judkins coupe was the first design executed by Gordon Buehrig after he became chief body designer for Duesenberg. It was a special order for Minneapolis industrialist Shreve Archer. The car was painted black originally. It survives today and is painted two shades of blue and it sold for $95,000.

Judkins Victoria Coupe

My design sketch for the Judkins Victoria Coupe Duesenberg was made by the process described in Chapter Three. The date on the drawing is February 13, 1930.

The car used the short wheelbase chassis and achieved its unusual proportions by virtue of its unique interior. Some other cars built several years before had used a similar interior but the Judkins Victoria Coupe was the only Duesenberg with this plan. It seated three people in comfort. By using the occasional folding seat, four could be crowded in.

This small personal coupe body on the Duesenberg chassis provided one of the most glamorous of all Duesenbergs in exterior

This side view photograph was taken by Judkins after they had completed building the victoria coupe to Gordon Buehrig's specifications. Only two of Gordon Buehrig's Judkins victoria coupe design were built. They were stark in appearance, with only the color panel below the door window openings for accent.

appearance. Its styling detail is severely plain and architectural with only the color panel in bas-relief below the windows to give an accent.

Only two were built, probably because the interior had a limited appeal.

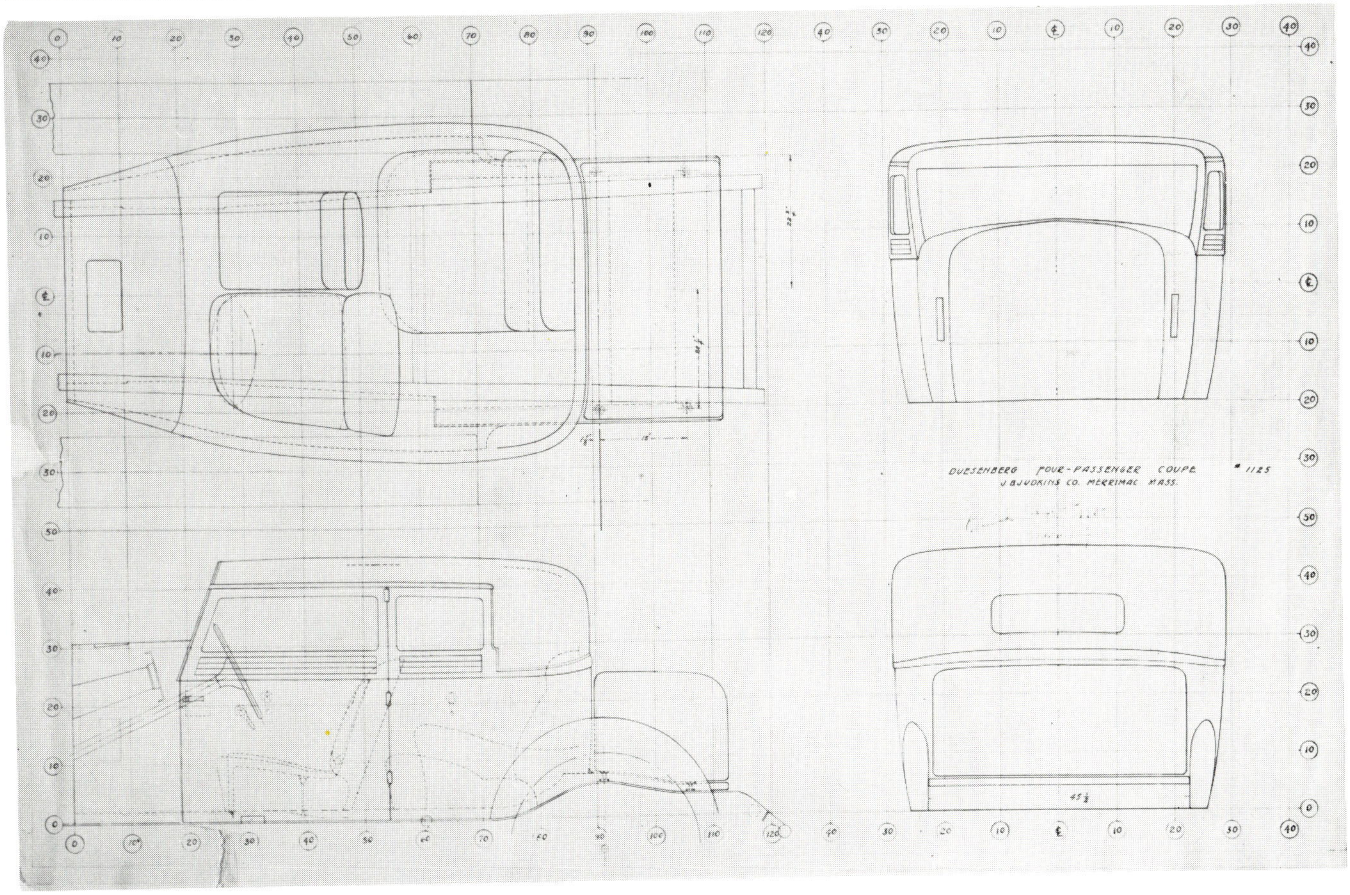

This designer's drawing of the Judkins victoria coupe shows the unique seating arrangement. It seated three in comfort and four when the occasional jump seat was folded down.

Beverly Berline

The first Duesenberg Beverly Berline was exhibited at the Auto Salon in Chicago on November 8-15, 1930. It was finished in three pastel shades of green. The fabric-covered top was a light tan Haartz cloth and the interior was a fawn colored broadcloth.

This design sketch is dated September 12, 1929, and indicates the car is to be built on the long chassis.

For the second body I selected three shades of tan for the exterior. For the interior I selected a rich maroon color chip and specified the color for the broadcloth interior trim and to have it piped in a light tan leather matching the light tan color stripe on the belt molding.

The dark interior was such a daring innovation that it was questioned by the vendor. When our purchasing department sent the maroon color chip to the Laidlaw Company and ordered a few yards of broadcloth to match it, we received a very carefully worded letter in return respectfully suggesting we obviously had made a mistake. The letter stated the color chip must represent the exterior color of the car and we wished a broadcloth which would be appropriate for the interior, but certainly we didn't mean to have the cloth in maroon. At our insistance, they

This body builder's drawing of the Beverly Berline shows the extra detail Gordon Buehrig put into his designs. He specified dimensions in all three planes and even (See Section A-A) specified construction in what might be a questionable area.

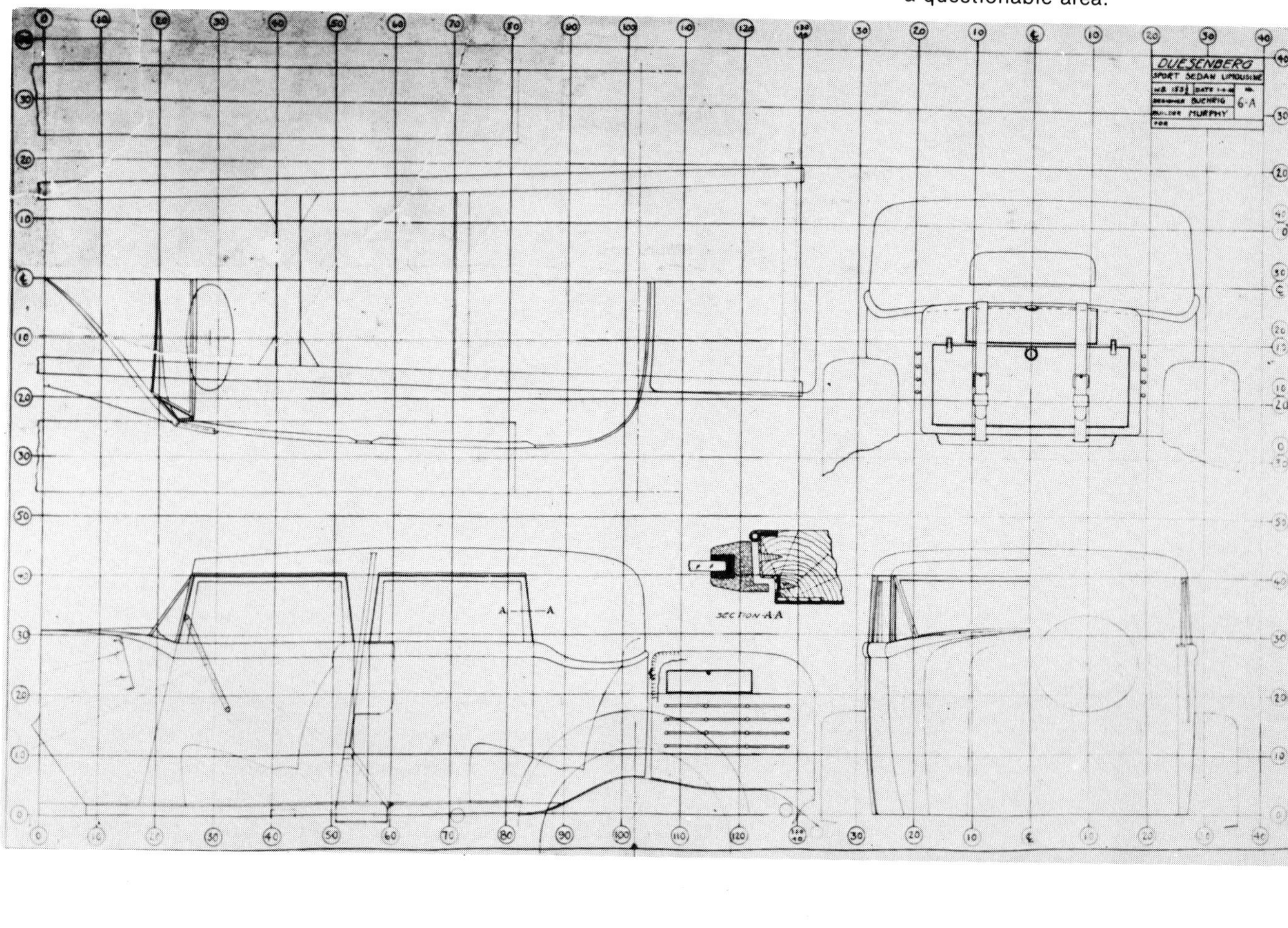

produced what we had ordered and the resulting interior was sensational. A few years later I used the same interior on the Model 810 Cord. More recently, John Reinhart used it for the Mark II Lincoln Continentals.

One of the design characteristics of the Beverly is the way the top material extends down to the belt between the front and rear doors. This detail in a slightly different form was mentioned earlier on the design model of the 1929 Buick, but was not used on the production car. The real origin of it goes back to the convertible four-door bodies built by Hibbard and Darrin of Paris.

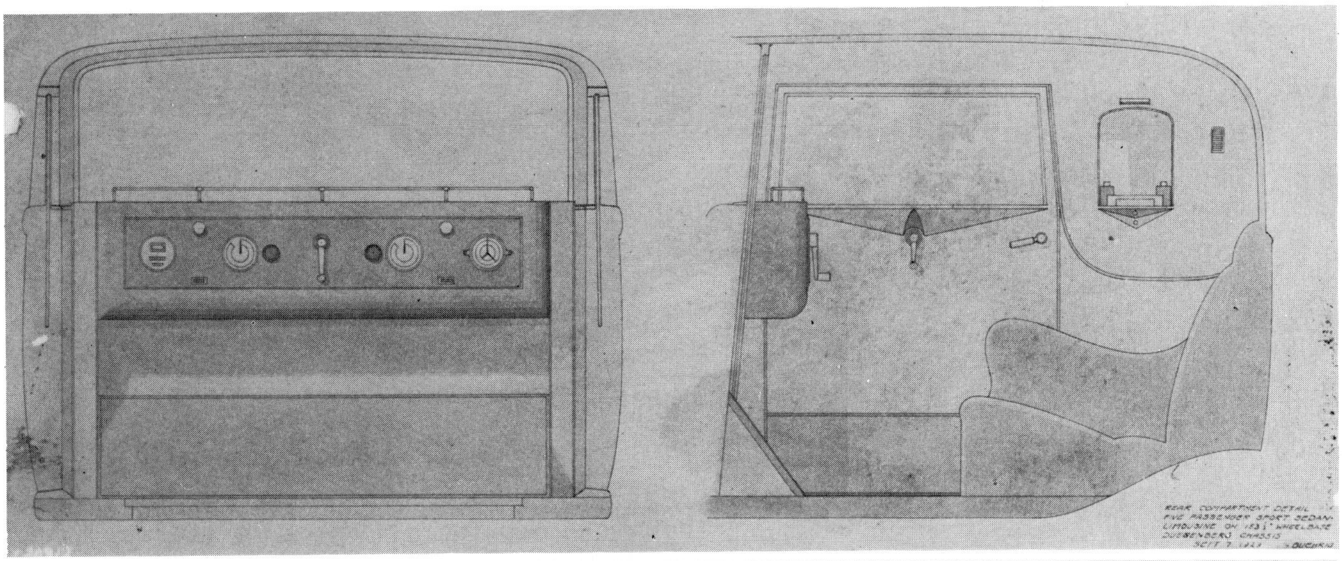

Buehrig's specifications didn't stop with the exterior. This sketch, dated September 7, 1929, details the interior appointments for the back seat of the Beverly Berline.

The finished interior of the Beverly Berline closely followed Gordon Buehrig's sketch. Note the addition of a rear seat speedometer.

One of the basic engineering problems of a four-door convertible is what to do with the space between the front door and the rear door windows. Since the doors cannot have structure above the belt and the glass run channels are inside the pillars, there is a void of three or four inches between the front and rear door glass when they are in the raised position. A common solution was to have a removable pillar fitted in or, in some cases, to have a pillar hinged at the bottom to fold inward. Tom Hibbard and Howard Darrin solved the problem by having the top material extend down to the belt between the windows. This earflap effect was unique to Hibbard and Darrin bodies but was copied, probably with permission, by Derham on one or two of their custom bodies. On the Beverly Berline the effect in this area was crisp and geometric, an

This builder's photograph shows a new Beverly Berline in what appears to be a New York City Central Park setting. As Beverly Berlines were built by more than one body company, this is possible.

This rear 3/4 view of a new Beverly Berline shows the body had good proportions from almost any angle.

effect I could accomplish since it was not a convertible and because I used cast metal upper door frames.

The windshield on the Beverly Berline was a purely Buehrig invention. It allowed a greater slant on the main glass panel and brought the "A" post back to allow a straight line door opening in a reasonable position. The windshield header was positioned farther back on the body by virtue of the arrangement and this effect gave the vehicle more dramatic proportions.

The Beverly Duesenberg was a Berline (pronounced Ber-len). It refers to a small limousine which does not have folding jump seats in the rear and which has broadcloth trim in both the front and rear seats. Consequently, it serves as an owner-driven sedan, yet it is equipped with a glass partition behind the front seat and may be used as a chauffer-driven vehicle. In contrast, the typical limousine has jump seats in the rear and therefore is a larger body. The front is trimmed in leather for the chauffer and the rear in broadcloth and is not intended to be driven by the owner.

In the cabinet behind the front seat we installed a radio, a glove box and a rear seat instrument panel. Since the Duesenberg Model J was the only American car capable of 100 miles per hour, I reasoned the rear seat passenger would enjoy a speedometer.

The large fitted trunk on the Beverly made it a practical car for touring. I am very proud of this design and it is a source of personal satisfaction to hear about one of these cars, now over forty years old, selling for $100,000 to $150,000. The original price was $14,000.

LeBaron Phaeton - designed by LeBaron

1930 Duesenberg Beverly Berline - a Buehrig design

DUESENBERG - 'SJ' (Short w

lbase) - Cabin Speedster

Duesenberg La Grande Dual Cowl Phaeton.

The supercharged Duesenberg engine.

My personal favorite Duesenberg is the Derham Tourster. I designed this car for the 1930 Salon. It is severely plain in ornamentation and has the unusual virtue of being equally handsome with the top in the raised position or when it is lowered.

One detail designers tried to achieve in phaeton design was a horizontal and low stacking of the top in the lowered position. This was sometimes achieved by using very light quarter padding in the top, resulting in a rather bony effect when the top was in the raised position, and by building the top wide enough to allow the bows to fold outside the body where it was widest to accommodate three passengers in the rear seat. The LeBaron phaeton was made this way and was much more attractive with the top down than in the raised position. Both are beautiful cars but I think the top on the Tourster is an improvement over the Ralph Roberts' design.

The LeBaron phaeton was produced on both the long and the short wheelbase chassis. On the short one the rear seat was positioned over the rear axle. The Derham Tourster was designed only for the long wheelbase chassis which positioned the rear seat forward of the axle. Therefore the rear seat could be lower to the ground. This advantage allowed for a better roof profile in side elevation and still have adequate rear seat headroom.

Gordon Buehrig's personal favorite design was his Derham Tourister on the Model J Duesenberg long chassis. The first one was painted a wild combination of pale green fenders and a vivid yellow body. It was the show car and when it appeared in Los Angeles, movie actor Gary Cooper, seen here in the driver's seat, walked in, saw it, and bought it!

Below: One of the attractive features of the Derham Tourister is that it looks as good with top up as with top down—something that wasn't always true of touring cars of the day.

The front seat of the Derham Tourster was designed to be wider than on the LeBaron phaeton and the rear seat was more narrow. This made the sides of the body more parallel in plan view. The wider cowl gave a wider base for the windshield. The top width at the front has to be wider than the body at the rear if it is to have a low stack height. Since the Tourster is more narrow at the rear seat, the top is also more narrow. This results in a windshield which has pillars that are more parallel when viewed from the front of the car. This is a very desirable styling effect.

All of this is accomplished in the empirical design lines in plan view on the body draft. The width of the body at the front seat determines how much reverse curve must be put in the cowl to blend the body into the hood plan view lines. This reverse curve is called an ogee line. A body engineer would express it this way: To widen the front seat you would put more ogee in the cowl.

A popular feature of expensive phaetons was a tonneau windshield. This was usually a folding shield attached to folding cowl which was pivoted at the back of the front seat. This device did a good job of protecting the rear seat passengers from the wind when the top was down and worked reasonably well when the top was up. However, it was an abominable gadget when the side curtains were fitted. In order to get into the rear seat you first released the friction knobs at each side of the screen and folded it forward. Then you reached under the cowl and found the handle which released the locks on either side so the cowl could be raised. After you were seated, you repeated these steps in reverse order and did this every time you used the rear seat. Of course, it did look great and did provide comfort for the rear seat passengers, but at a cost of inconvenience.

On the Derham Tourster the same result was accomplished and all the nuisance factors of the dual cowl body were overcome by inventing a crank-down tonneau windshield. This feature was so well liked it was copied later by Cadillac and Chrysler. The crankhandle, similar to a regular door window crank, was mounted on the back of the front seat in a recessed well.

By using footwells between the side rails of the frame it was possible to move the rear seat forward and still have adequate rear legroom. Since the body was designed for the long wheelbase chassis there was adequate rear door width to allow a fixed cowling eight inches deep to extend into the rear compartment and still have easy access to the rear seat. This allowed positioning of the glass to run on channels set at a 35 degree angle off the vertical. The result of this was the relative position of the tonneau windshield to the rear seat passengers was almost identical to the regular windshield in its relative position to the front seat passengers.

On the Cadillac copy of the idea the tonneau windshield is so far from the rear seat passengers that wind protection is not effective. It also looks somewhat out of place.

There was an amusing mistake made on the first Derham Tourster. It was to be the show car and I selected a pale green for the fenders, belt molding and interior. Even the steering wheel was in this color. For the body I selected a duPont color called Goldenrod Yellow. It was a subdued color which matched up well with the green. When the body arrived from Derham I was shocked to see it in a strong vivid yellow and not like the color chip which had been selected. It seems the color chip was for an obsolete duPont color and the same name had been used for the bright yellow.

Both Ames and I were a little stunned but we decided to buck the body (put it on the chassis) and finish the car anyway, deciding if it looked too bad we would repaint it.

It was a wild combination but not offensive and Ames decided it might make a "Show Stopper." From the show at the Drake Hotel in

Chicago it was shipped to the Salon in Los Angeles. Movie actor Gary Cooper walked in, saw it and bought it.

Today, Toursters are among the most desirable of the Model J Duesenbergs, and trade in a price range of $60,000 to $100,000 . . . that is, if you can find a seller.

Torpedo Phaeton

The most unusual design assignment I had at Duesenberg was for Mark Lawrence. It all started one day when Ames asked me to come down to his office. He handed me an eight-page letter written in longhand and containing a few crude sketches. He said, "Gordon, here is a letter from a boy in Washington, D.C. who says he wants to buy a Duesenberg built to his own ideas. He has a Model A Ford with racing equipment on the engine that he wishes to trade in on the Duesenberg. It is obviously from an automobile-crazy kid and I would like to have you answer with a nice letter. Send him a sketch or two if you have some and a catalogue."

I took the letter upstairs and read it several times, then went back to see Ames. I told him I agreed we couldn't consider this young boy a serious prospect, but was impressed with his original ideas. I said, "With your permission, I would like to work out a design around Mark Lawrence's letter and it might be something we would want to build for an auto show." Ames agreed.

When I showed the finished sketch of my proposed design to Ames, he liked it and told me to send it on to the boy in Washington.

It turned out Mark Lawrence wasn't a boy but a very nice young man about twenty years old. When he received the sketch he called Ames and said I had designed exactly what he wanted and how much would it cost?

Our Duesenbergs were in the general price range of $12,000 to $14,000, but these prices reflected a production of five or ten cars of the same design, but with custom trim and color. A one-of-a-kind, which we thought this would be, cost more. Also, this body was quite complicated. Ames decided to have Brunn build it. After getting a quote on the body from them he set the price at $19,000 and informed Mark Lawrence. Mark wired back, "Proceed immediately."

This factory photograph shows the original Buehrig-designed Torpedo Phaeton which was built by Brunn for Mark Lawrence, the young man who originally suggested the car in a letter to Duesenberg.

Gordon Buehrig's design sketch of the Torpedo Phaeton was based on rough ideas in a letter received from Mark Lawrence, an automobile-crazy young man from Washington, D.C.

The exhaust side of the Mark Lawrence Torpedo Phaeton is seen here, with the welded steel exhaust manifold. Note the stone shield in front of the radiator shutters and the Duesenberg script on it.

These two views of the Torpedo Phaetons, one built by Weymann (top), the other by Walker-Le Grande (bottom), as photographed prior to delivery. Note the rather unique side curtain installed between the two windows to fill the area over the rear seat tonneau cover.

I finished my scale body draft, interior design layout, color and trim specifications and sent them on to Brunn. This model was later designated the Torpedo Phaeton and five of them were built. This included the first one by Brunn and two each by Walker-LeGrande and Weymann.

When the body for Mark Lawrence's car had been decked and all details of construction nearly finished he arrived in his Model A Ford but he wasn't driving it. His chauffer was. He also changed his mind about trading in the Ford. Instead, he gave it to his chauffer.

The day Mark took delivery he was probably the most excited customer who ever drove a new car from the plant. Mark had a tendency to stutter and his excitement added to his speech problem but it did not detract from the great grin on his face. He had a fist full of five dollar bills. He walked through the plant, shook hands with every workman, handed him a five dollar bill and thanked him for helping to build his Duesenberg.

We were indebted to Mark Lawrence in many ways but most important, his ideas had inspired one of our greatest models.

The Torpedo Phaeton had a folding windshield and a folding tonneau shield. I didn't want the Duesenberg to have the same fault as the folding windshield on my Buick roadster, which folded forward and was secured in the raised or folded position by friction from a hand-tightened threaded stud with a knurled knob. I knew from experience the folded windshield would often fly up from the lowered position at high speed because of wind pressure. It was a frightening experience.

On the Duesenberg design the pivot point was assembled with a bolt through the bushing on the stanchion and the shield member. About an inch below the pivot point a tapered pin bolt was located which screwed into one of the members and matched two notches in the other member. One of the notches located the windshield in the down position and the other one in the up position. No amount of wind pressure could move the shield from either position.

This view of the rear seat of the Mark Lawrence Torpedo Phaeton shows the instrument panel and the custom hardware.

Left: Interior door hardware similar to that used on the Laurence Torpedo Phaeton. Right: The handcrank for the Tonneau windshield of the Derham Toursters.

Brunn Town Car

The Brunn Town Car was the only town car I designed at Duesenberg which was built. It involved an invention on which we applied for a patent. The car was a success and the patented idea was good except for the fact that market for town cars had almost vanished. It was the last town car design built by Duesenberg. Ten years earlier the design would have been valuable.

Frankly, the town car was designed to emphasize class distinction. The chauffeur wore a uniform and was a servant to his master. Having the chauffeur's compartment open and the owner's enclosed emphasized the difference in status. The owner spoke to the chauffeur by using a microphone connected with a speaker mounted near the shoulder of the chauffeur.

The Brunn Town Car with white sidewall tires shows off the elegance of the design.

In automobile art, perfection may be classified in three categories: proportion, detail design and workmanship. Proportion is number one.

The purpose of the invention on the Brunn Town Car was one to achieve dramatic proportions. Probably the smartest looking town car at the time was Brewster's Rolls-Royce with the enclosed portion of the body containing only a bench-type seat for two people. Certainly it had beautiful proportions and snobbishness to the tenth power.

Historically, however, the bulk of the town car market was for the larger enclosed body with jump seats and capacity for five people. I decided to try to approach the proportions of the Brewster Rolls-Royce, yet seat five in the rear compartment.

We read the proportions of a town car as the ratio between the length of the enclosed portion of the body and the rest of the car. On the Brewster Rolls-Royce this ratio was quite large because of the small two-passenger enclosed rear portion of the body. On the Brunn Town Car a similar proportion was achieved by a trick interior layout which is shown on my original watercolor interior design layout. This interior sketch was made for another design which was not built. The physical layout is exactly the same as on the Brunn job.

As can be seen from the accompanying illustration, the interior seating for five, yet the maintenance of the exterior proportion similar to the Brewster Rolls-Royce seating only two was accomplished by putting part of the rear interior space—below the belt—ahead of the glass partition. The interior sketch shows how this was done. A cowling back

The occasional jump seats in the Brunn Town Car folded out of the panel at the front of the compartment and also uncovered an open area under the cowling where jump seat passengers could put their feet.

of the driver about ten inches deep covered the area inside the body where the jump seats were folded and concealed by a pull-down fabric curtain. When the jump seats were in use, those passengers feet and part of their legs were under the cowling. It was all very roomy and comfortable and the exterior proportions were dramatic.

Duesenberg management thought enough of the design to use it for the Duesenberg advertisement in the 1930 Auto Salon souvenir catalogue. Only one car was built. It was sold to Victor Emmanuel, a wealthy industrialist who lived in Ithaca, New York. Unfortunately, only the engine exists today, the body was scrapped in the 1950's.

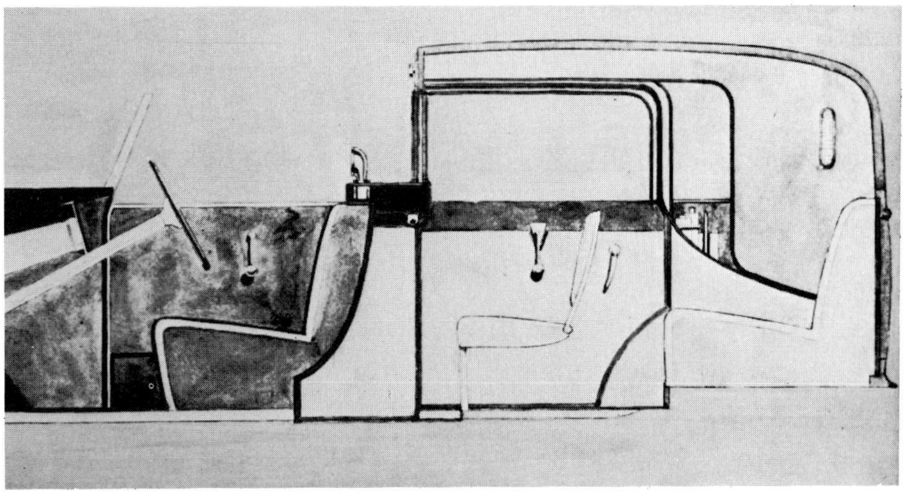

A design sketch by Gordon Buehrig shows how he was able to retain the dramatic proportions of the Brunn Town Car, yet provide seating for four persons.

The Twenty Grand

One of the idioms which developed during the Prohibition Era was the term "Grand" to designate $1,000. When Duesenberg first displayed a new model which they named the "Arlington" with a price tag of $20,000, it immediately picked up the nick-name of the "Twenty Grand."

I designed this car to be our exhibit at the 1933 World's Fair in Chicago. Only one was built and Rollston was the builder. The design is an attractive combination of a greenhouse reminiscent of the Beverly Berline, but with even more Hibbard and Darrin influence. The rear of the body is similar to the Torpedo Phaeton. The custom interior hardware was the same as I designed originally for the Mark Lawrence car.

I had a thing about hinges in those days and several of my Duesenberg designs had either several heavy chrome plated exposed hinges or a large piano type. Concealed hinges were available and were used on some of my designs, but I favored the exposed type because structurally they were more honest. The "Soss" concealed hinges required a huge cut out in the hinge pillar which reduced its strength. On the Twenty Grand there were fourteen exposed chrome plated hinges.

Original color of the Twenty Grand was grey with a grey leather interior. The top material was a very light tan Haartz cloth. As with the Beverly Berline, the Twenty Grand was equipped with a rear seat instrument panel.

I don't know the ownership history of this car, but it is now owned by Benjamin Johnson of Connersville, Indiana. The car is now painted black with a black top.

The Duesenberg SJ four-door sedan built by Rollston to a Gordon Buehrig design for the 1933 Chicago World's Fair exhibit was tagged the Arlington Sedan by the company. This quickly became the "Twenty Grand" when it was displayed with the $20,000 price tag.

This side of the Twenty Grand shows Gordon Buehrig's love for multihinges, plus his admitted borrowing again from Hibbard and Darrin in the vee of material coming down between the doors.

The interior of the Twenty Grand sedan featured the same hardware as the earlier Torpedo Phaeton, plus the individual rear seating of the Beverly Berline. Note it also had a rear seat instrument panel.

Rollston Convertible Torpedo Victoria

Early in 1932, the Duesenberg factory scrapped the body of a Willoughby sedan and had Rollston construct what we called a Convertible Torpedo Victoria on the chassis, which carried number J-231. It is now owned by Terence E. Adderley of Bloomfield Hills, Michigan, and is in the process of being completely restored.

The car has beautiful, yet unusual proportions. The philosophy behind the design was that in order to give easy access to the rear seat,

The Rollston Convertible Torpedo Victoria was built by Rollston based on a Gordon Buehrig design for the 1930 Chicago Salon. The doors were so long that hardware to open them was provided at each end of the door.

cut down on the blind quarter area usually associated with convertible victorias and emphasize the longitudinal effect of the window area. As a result, this car had unusually long doors. On this I used the huge chrome plated piano hinges to good advantage.

Because the doors were extra long it would have been difficult to release the door locks from the rear seat, so we used two remote control handles on each door to release the door locks. The most attractive interior was the design work of J. Herbert Newport.

The attractive interior of the Rollston Convertible Victoria was the design work of J. Herbert Newport. Note the dual door hardware. . . and aid to rear seat passengers.

The Whittell Speedster

The Whittell Speedster, now in the Harrah Automobile Collection in Reno, Nevada, is one of two boattail bodied Duesenbergs I designed. Both were built by Burt Walker at his Weymann body company in Indianapolis. The contours of both bodies are similar except one is on the short wheelbase chassis and the other is on the long wheelbase chassis.

One of the Speedsters had crank-down windows in the doors and the other was fitted with side curtains in true roadster fashion. In my

The 1932 Duesenberg Model SJ with Weymann-built "Fish Tail" speedster body by Gordon Buehrig. Note the "Duesenbird" radiator ornament motif is carried over into a similar spear on the side just behind the door—an idea Buehrig carried forward into the Auburn 851 Speedster.

opinion, the car with the 142.5-inch wheelbase was a better proportioned vehicle. However, Mr. Whittell specified the long chassis for his Speedster.

These two cars are extremely interesting in that they are similar to the Auburn Speedster I designed a couple of years later. My interest lies in the fact I designed the Duesenbergs with one method and used another to do the Auburn. The subtle differences in body shape between the Auburn and Duesenbergs are a result of the methods used in designing them. This is described in detail in the chapter dealing with the Auburn Speedster.

George Whittell of San Francisco was Duesenberg's best customer in respect to the number of cars he owned. His stable of cars included six Model J Duesenbergs.

The Whittell Speedster was acquired and restored by Harrah's Automobile Collection, Reno, Nevada, and looks this way today.

As can be seen here, the Whittell Speedster had a top and windows. It was a more attractive car with the top down.

Derham Four-door Convertible

The Derham Four-Door Convertible has several typical design characteristics of mine. These include the heavy, exposed plated front door hinges and the piano-type hinges on the rear door. The trunk design is also the same as I used on the LaGrande Phaeton. It is a leather-covered trunk with chrome plated trim which emphasizes the development of the trunk shape.

The rear compartment instrument panel is equipped with a chronograph and a speedometer. The car was designed for the short wheelbase chassis because long wheelbase frames were in short supply.

LaGrande Phaeton

The LaGrande Phaeton design sketch was originally created at Harole Ames' request for Gene Markey of Beverly Hills, California. Gene Markey was the husband of movie start Joan Bennet. The design is obviously a copy of Ralph Roberts' LeBaron design.

The Derham Four-Door Convertible, or convertible sedan made use of the short chassis and had several Buehrig trademarks—the heavy piano door hinges which were chromeplated, and the trunk. It was a New York Salon car in 1932.

This interior view of the Derham Four-Door Convertible shows the rear compartment instruments.

The LaGrande Phaeton was a product of the depression and it exemplifies the struggle Harold Ames made to keep the Duesenberg Company in business.

The Union City Body Company of Union City, Indiana, who built the body shells for the LaGrande Phaeton, was not a prestige carriage builder. They had built some bodies for Auburn, but they also built seats for movie theaters and other non-automotive products. However, they could build bodies and they had a low wage scale. Ames was able to buy phaeton bodies from them for less than he could from LeBaron or Derham.

At the Duesenberg factory we had good trimmers and painters, equal in fact to those in the prestige custom shops. Thus, the LaGrande bodies were purchased "in the white," then decked, trimmed and painted at Duesenberg. This arrangement had the extra advantage of keeping our own craftsmen busy.

Most of the LaGrande bodies were mounted on the short wheelbase chassis because, as mentioned, we had more of them.

LaGrande, incidently, was a name coined by Harold Ames to be a prestige body nameplate because Union City Body Company was not a recognized carriage builder. The LaGrande name was also used later for all bodies received at the factory "in the white," which included unfinished bodies from Weymann, Walker, Brunn and others. While a LaGrande body plate was struck, it was not normally used, and these vehicles were usually delivered with no body plate affixed, as they were referred to as Duesenberg's own coachbuilder. They are credited with producing twenty-one bodies, of which nineteen were LaGrande Phaetons.

Some of the improvements in the proportions of the body I was able to make on the Derham Tourster over the LeBaron Phaeton were also carried over to the LaGrand Phaeton.

The short wheelbase limitations excluded the crank-down tonneau windshield layout. The extra cost of this innovation also ruled it out. The standard old fashioned tonneau cowl could be sold as an option.

I put more ogee in the cowl and made the body more narrow in the rear (in comparison to the LeBaron design). By doing this, the top design could be made very similar to the one on the Derham Tourster.

A factory photograph of the Buehrig-designed La Grande "Sweep Panel" Phaeton. This was a dual cowl phaeton on the short wheelbase chassis.

A late La Grande Dual Cowl Phaeton finished conservatively and with the optional full wheel covers over the wire wheels.

The windshield design from the Torpedo Phaeton, including the positive lock for the folding windshield, was also incorporated in the LaGrande Phaeton design.

On the short wheelbase LaGrande Phaeton, the rear of the body extended so far aft of the rear axle that only a small trunk could be fitted. A large trunk was desirable to improve the proportions of the car, so I designed a higher one to give the trunk area more mass.

At the forward top area a folding false compartment was included in the design which provided space for the top stack to rest in a desirable low position with the top down. When the top was in the raised position the false compartment folded out to give the car improved proportions and increased mass.

The Duesenberg radiator ornament

Earlier I explained the chassis sheet metal and all the chassis accessories, including lamps, bumpers, etc., were designed before I came to Duesenberg and I never redesigned any of them. There was one item in this area which I did design because it was added later. This was the Duesenberg Radiator Ornament.

Harold Ames had a unique idea about the car at its inception. He felt the Duesenberg would be so distinctive that a nameplate and a radiator ornament would be superfluous. Owners, however, became weary of telling people their car was a Duesenberg and several asked to have a nameplate added. Many of them also wanted a radiator ornament and they bought them in all shapes and sizes and many looked terrible on the Duesenbergs. Some of them were very expensive, like the Lalique glass ornaments from France. Some were purchased in jewelry stores.

None of the ornaments we saw seemed to have the proper character for a Duesenberg. So Ames asked me to design one which would enhance the car.

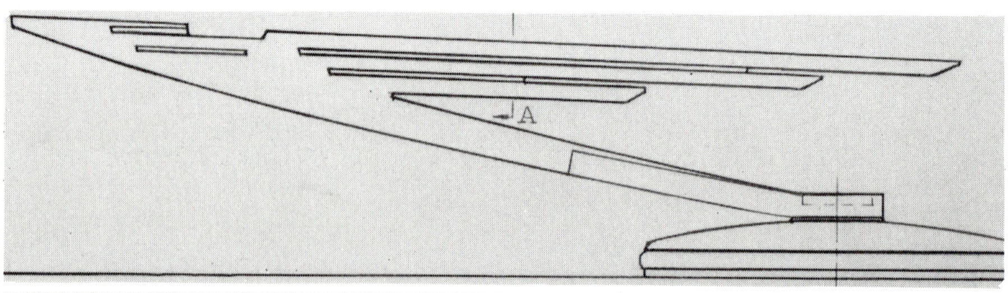

This sketch shows the 2/3 scale Duesenberg radiator ornament Gordon Buehrig did for a trophy for the Auburn-Cord Duesenberg Club. In creating the original design, Buehrig comments he worked out something which could be built primarily on the machine tools in the Duesenberg shop as opposed to expensive tooling which many radiator ornaments required. Duesenberg had men and machines, but few dollars to expend for tooling.

The Duesenberg radiator ornament, created by Gordon Buehrig in June of 1929 when Duesenberg gave in to owner demand for such an identifying item.

Duesenberg's unique economic situation had a bearing on how I designed it. They had a labor vs. tooling aspect which was the exact opposite of the high production automobile companies. The general rule in the automobile business is the more money you invest in tooling, the less money you need for labor. We were building less than a hundred car a year and our cars were expensive. We could afford labor but we couldn't afford high tooling costs.

I was aware of this and designed the ornament so it could be made in the Duesenberg shop without special tools. It is made from two flat sheets of brass, cut to profile. A few lines are cut in one of them on a milling machine. A hole is drilled in the base piece and then the two pieces are brazed together and chrome plated. The highly stylized bird in flight complimented the beautiful lines of the car.

A few years ago Harold Ames wrote me and said he would like to donate a suitable trophy to the Auburn-Cord-Duesenberg Club. I suggested a smaller version of the Duesenberg Radiator Ornament be used. I made a two-thirds scale drawing of it and sent it to Ames in Chicago where he had it made by C.D. Peacock in Sterling Silver. The Harold Ames trophy is a perpetual award given each year to the "Best of Show" at the meet in Auburn, Indiana.

A Roadster design by Gordon Buehrig on the Duesenberg Model J short chassis which was never built.

A Sport Phaeton design by Gordon Buehrig on the 153 1/2 inch wheelbase which was never built. Note the cut down doors.

Additional pictures of Buehrig designs

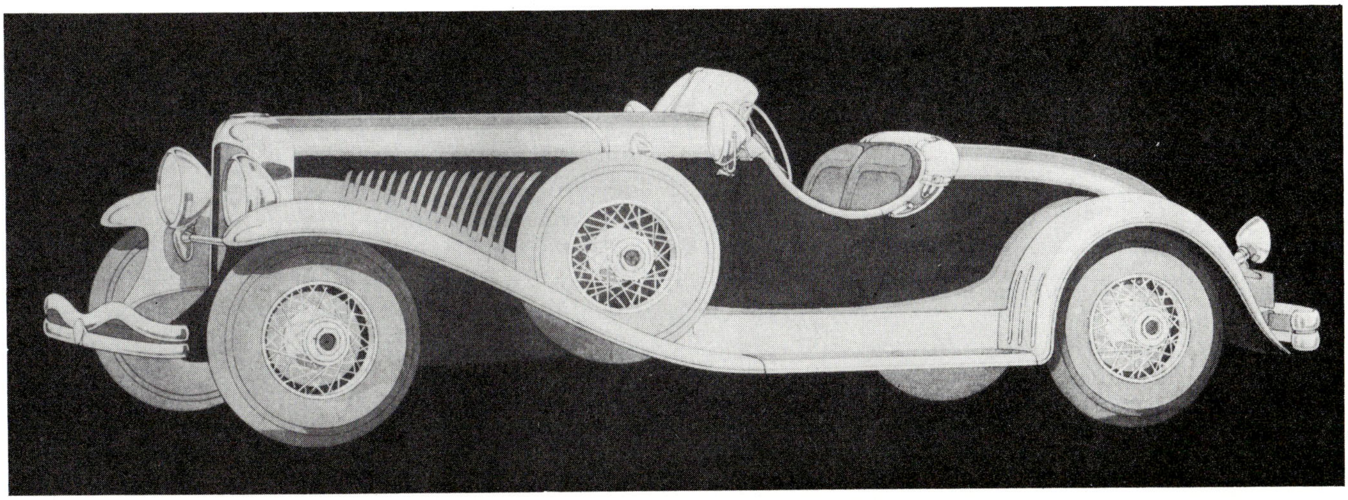

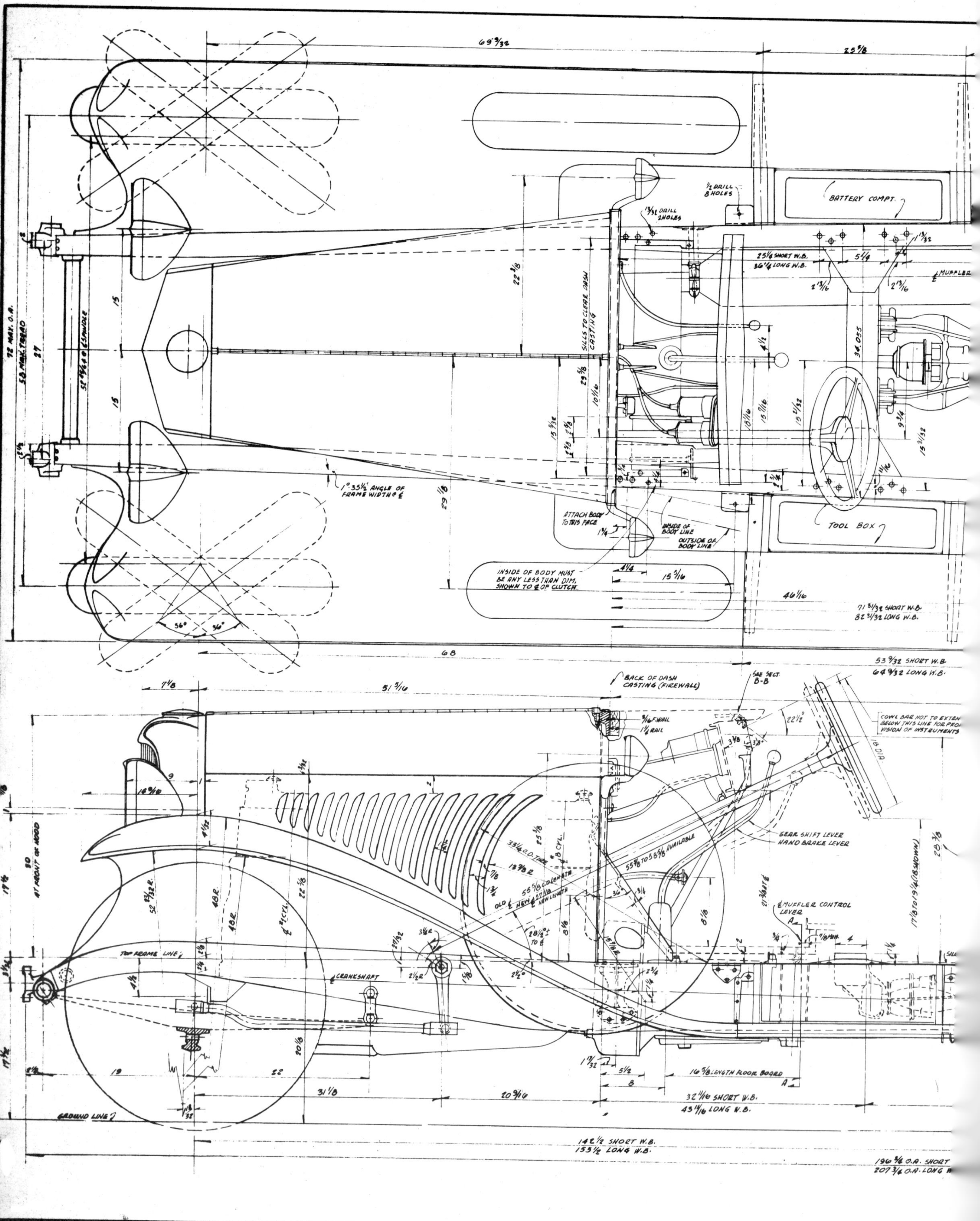

69 9/32
25 5/8
BATTERY COMPT.
TOOL BOX
1/2 DRILL 8 HOLES
13/32 DRILL 2 HOLES
25 1/4 SHORT W.B.
36 1/4 LONG W.B.
MUFFLER
SILLS TO CLEAR DASH CASTING
1° 35 1/2 ANGLE OF FRAME WIDTH @ ℄
ATTACH BODY TO THIS FACE
INSIDE OF BODY LINE
OUTSIDE OF BODY LINE
INSIDE OF BODY MUST BE ANY LESS THAN DIM. SHOWN TO ℄ OF CLUTCH
72 MAX. O.A.
58 MAX. TREAD
46 1/16
71 31/32 SHORT W.B.
82 31/32 LONG W.B.
68
53 9/32 SHORT W.B.
64 9/32 LONG W.B.
BACK OF DASH CASTING (FIREWALL)
SEE SECT. B-B
COWL BAR NOT TO EXTEND BELOW THIS LINE FOR PROVISION OF INSTRUMENTS
18 DIA.
GEAR SHIFT LEVER
HAND BRAKE LEVER
℄ MUFFLER CONTROL LEVER
AT FRONT OF HOOD
TOP FRAME LINE
CRANKSHAFT
GROUND LINE
16 5/8 LENGTH FLOOR BOARD
32 11/16 SHORT W.B.
43 11/16 LONG W.B.
142 1/2 SHORT W.B.
153 1/2 LONG W.B.
196 3/4 O.A. SHORT
207 3/4 O.A. LONG

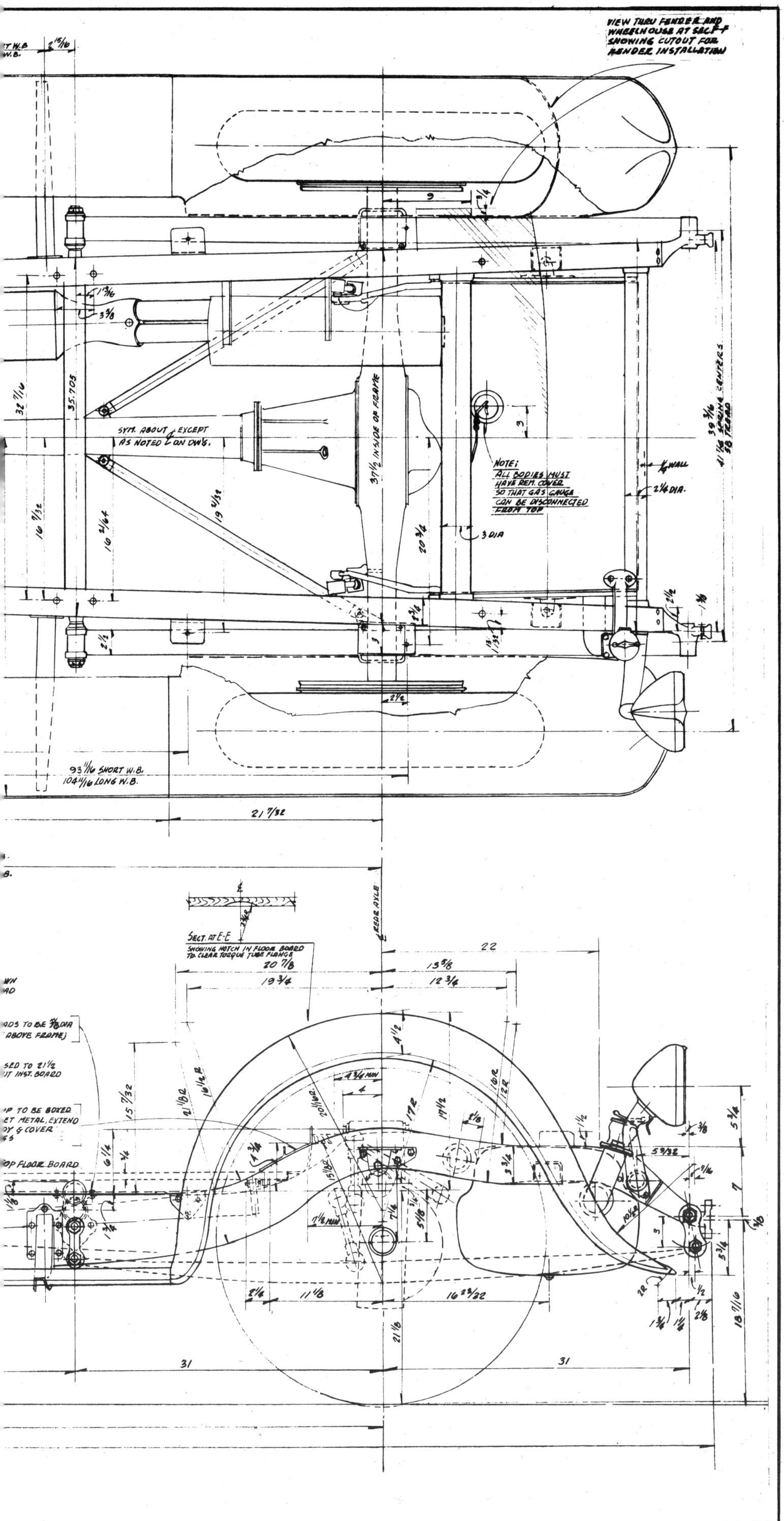

Above is the Duesenberg Model J Body Builder's Chassis Drawing. It was worked out by the engineering department at Duesenberg and sent to all custom body builders. It contained all the information a body builder would require in fitting custom coachwork to a Duesenberg Model J chassis.

A versatile body design by Gordon Buehrig.

One of the first designs done for Duesenberg by Gordon Buehrig after his arrival was this phaeton built by LeBaron to his design. It was exhibited at the Chicago Salon in November 1929 and was the first Duesenberg with a Buehrig body which was exhibited. It is seen here ready for shipment on Le Baron's loading dock. The car was black, trimmed in red leather. Two chrome belt moldings framed a color accent panel in red.

This view of the Judkins Victoria shows its unusual seating arrangement.

This Beverly Berline is equipped with non-standard fenders which make the car look heavier. Designer Buehrig never approved them.

This Murphy-built Beverly with the standard fenders has a much lighter appearance.

Fred Duesenberg is seen here seated in a La Grande Dual Cowl Phaeton mounted on the SJ chassis with the race car type exhaust manifold. The picture was taken at the Indianapolis Motor Speedway.

Interior of the rear seat of the Walker-La Grande Torpedo Phaeton—the Mark Lawrence inspired car by Gordon Buehrig. Note speedometer, clock and compass.

Rear view of the Walker-Le Grande Torpedo Phaeton on the Duesenberg Model SJ chassis is hard to distinguish from the rear. The sloping deck makes the car reminiscent of a Murphy convertible.

This Duesenberg Model J Limousine was designed by Judkins for Duesenberg, but it carries a Gordon Buehrig-designed interior.

This interior shot shows the rear seat of the Duesenberg Model J Judkins Limousine with jump seat up. Note the characteristic Buehrig central window crank handle on the door.

When the car which is shown as a design sketch on page 54 was built, the body builder located the landeau iron, or top prop iron, outside the quarter panel. It was inside on the sketch.

The Buehrig Model A Ford | 5

During the years 1928 through 1931 millions of Model A Fords were built by the Ford Motor Company and I honestly think I had the most beautiful one of all. Mine was something a little different . . . it had a body by Buehrig.

I do have my original design sketch of what I expected it to be when I started to build it and, although I changed three small details when I built it, the original concept was not changed. The standard Ford trunk rack accessory folded to a vertical position and the permanent trunk ahead of it was changed in design to have a vertical rear face. The other change was to eliminate the outside landau iron and put it inside. I also eliminated the Hispano Suiza radiator ornament and put on a "more honest" Motometer.

This original design drawing was lost for over thirty years. At an Auburn-Cord-Duesenberg meet one year, a man from Indianapolis introduced himself and told me he had a present for me from our mutual friend, Oscar Hadley. Oscar was the blueprint boy and file clerk when I started at Duesenberg and later he became a draftsman and then an engineer. When the Duesenberg plant was being closed down, Oscar retreived my Model A original design sketch from a wastebasket and kept it for me all these years. I am indebted to him to have it and be able to present it here.

On September 27, 1930, I took delivery of my 1930 Model A Ford cabriolet, complete with six wheels, trunk rack, rumble seat and special eight-ply General white sidewall tires and drove it about four miles from the dealer to Augie Duesenberg's race car shop, where the next day Augie's men went to work on it.

This is Gordon Buehrig's original design sketch for his Model A Ford cabriolet conversion. Note the Hispano Suiza stork on the radiator.

First we removed the body and cut it in half immediately behind the "B" pillar and removed the convertible top. The rear half of the body, which included the rumble seat, was discarded. The front seat was also discarded. All of these operations were called for on the full-sized body draft I had taken to Augie's shop.

The windshield was cut down three inches in height and the door hinge pillar was also cut in height by the same amount. The windshield was vertical, which allowed the door hinge pillar to extend above the belt and to have a hinge at the top. This automatically took care of reducing the window area height by the stop at the top of the door hinge pillar.

"Dash" is the sheet metal on the body which separates it from the engine compartment. Sometimes it is called the firewall, because if the engine catches fire it prevents the fire from reaching the body. The lower part of the dash and the toe board may be the same metal stamping or, as on the Duesenberg, the same metal casting. Thus, for definition here, "Dash" means the front of the body.

Without getting into too much body engineering detail, suffice to say the dash and toeboard panel was cut from the body and relocated four and a half inches forward (relative to the body) and the void filled by welding in a four and a half inch strip of metal. When the body was remounted on the chassis, the relocated dash and toeboard were in their original location and the body was four and a half inches aft of its original position. A special enigne hood was made which was four and a half inches longer than the original one, which was discarded.

Next step was to build the framework for the rear of the body. The full-sized draft was made on a drawing table 40 inches by 72 inches. Since the body was bigger than this, it required shifting section details to such an extent that the patternmaker who cut out the parts couldn't read the drawing. I had him mill the stock into wood blocks the correct size and I detailed the lines on each piece. The framework went together perfectly and Augie's race car body builder hammered out the aluminium skin surface to fit. He did a beautiful job.

The only other work to finish was parts for the folding top. I retained the folding "B" pillar by cutting three inches from the bottom of it. The linkage between it and the windshield header was also used. The remainder of the folding top bows and linkage had to be made according to my body draft.

When this work was finished, I paid Augie's bill and moved the car over to the Duesenberg plant for paint and trim. Here again, Harold Ames was very kind, and allowed their best trimmer and their best painter to do the work. This was after regular working hours and the company didn't charge me a thing. I paid the workmen their regular hourly rate.

The materials used in trimming the Model A were the same quality we used on the Duesenbergs. I did get a special buy on four hides of leather from the Weymann Body Company. The company was liquidating some inventory and I bought one hide of pigskin grain cowhide which we used to cover the trunk and three hides of an antique warm gray leather in a mottled pattern for five dollars a hide. These were for interior trim.

The Buehrig Model A Ford as it looked upon completion in its first form. Date: December 21, 1930.

A Model A cabriolet in its normal form as delivered by the factory.

The exterior was in a color called Capucine. It was a duPont lacquer and I was told it came from the color of the robes worn by the Capuchine Monks. It came in three shades: Dark Capucine, Light Capucine and Flame Capucine. (The spelling "Capucine" is duPont's.)

The body was painted Light Capucine; the fenders and belt molding Dark Capucine and Flame Capucine was used for the wheels and for striping. Metal parts and the steering wheel were in Light Capucine.

We used Marshall Knockland design seat cushions on the Duesenbergs and I designed a special set for the Model A Ford. Since the package layout called for a seat cushion height two inches lower than standard, the problem was how to make it have the luxury of a deep cushion. This type of seat is made up of individually packaged coil springs about one and three quarter inches in diameter. These are laced together in a metal frame. I achieved the low seating by having the row of coils around the perimeter of the seat cushion extend to the floor level and all the springs inside the outside row extend one and one half inches below the floor level.

The seat back was in two parts but they were not concave as with bucket seats. The driver's seat back was in a fixed position, while the right hand seat back would fold down to provide access to the rear compartment. There was one seat in the rear which seated the passenger in a sideways position. The rear seat cushion also extended below the floor line except at the front of the cushion, where it was notched upward from the bottom side for drive line clearance. The foot well extended well below the normal floor line and gave a very comfortable seating position. It was a little difficult to enter but Fred Duesenberg, who was over 50 and troubled with arthritis, rode in it and liked it. For young people it was not a problem.

The car was completed December 21, 1930, and it was christened "Baby." The first day I drove it down town to go to lunch I parked it in front of the L.S. Ayres Company. After lunch, when I returned, there was a crowd around the car extending into the street to the streetcar tracks. A policeman was trying to clear away the crowd. He was happy to see me arrive and solve his dilemma.

There is nothing I have ever owned which gave me as much pleasure, as much satisfaction or that I owned with greater pride than this beautiful little car. Its quality in every respect was equal to a Duesenberg and that was pretty good!

After driving the car a little over a year, some changes were made and the results were again very satisfying. Experimental baloon tires had just come on the market in a variety of sizes and I acquired a set from

This view shows the Buehrig Model A Ford as modified after approximately one year of use. Most visible changes were the addition of a set of experimental Goodyear balloon tires and wheels, and redesigning of the radiator to give it a configuration similar to a Hispano Suiza.

Goodyear with special wheels in about 15 x 6.50 size. The original tires and wheels were 19 x 5.00. Using the larger tires required wider front fender wells for the spare tires.

Dick McCarthy of Chicago had given me a speed conversion kit made by Harry Miller, the race car builder from Los Angeles. I had earlier designed a race car body for him. This changed the engine to overhead valves and boosted the compression.

The cast aluminium valve cover was a sand casting with Miller's name embossed on the top. I had this ground off and then the entire cover finished and polished. Other engine room improvements in this major operation included chrome plating the fan and all gas and oil lines. I added Ray Day aluminium pistons, Mallory ignition and special Packard Electric cable wiring.

I designed a new instrument panel made of one quarter inch thick aluminium sheet which was engine turned on the front face. Front mounted special gauges were made up special by the Pioneer Instrument Company, using their aircraft instrument cases. Direct lighting was common at that time and the two instrument panel lights were Pontiac. The panel extended the entire width of the cowl and was aft of the standard location.

The original gasoline gauge was retained. On the Model A Ford, it was located in the tank, which was also the top of the cowl. A chrome plated cylinder, polished on the inside to carry a beam of light from the instrument panel to the gauge, was used.

I redesigned the radiator to make it more the configuration of a Hispano Suiza and added a Motometer to the radiator cap. The area at the bottom of the radiator was housed in sheet metal with a scrapped Duesenberg gasoline tank cap added as a crank hole cover.

The top was redone with new Haartz cloth and the car completely repainted, this time in a rich red maroon. Bright red was used for striping.

The car was written up in MoToR Magazine at the time. I drove it 89,000 miles before selling it in 1934.

The Baby Duesenberg | 6

Strangely enough, the story of the Baby Duesenberg project started with my leaving the company for a brief sojourn with General Motors.

By the beginning of 1933, the outlook at Duesenberg was growing dim and I decided I should affiliate with a stronger company. I wrote to General Motors Styling and was invited back.

Back in Detroit, evidence of the Great Depression was far more apparent than it had been in Indianapolis. The auto industry was in very bad shape and, reflecting on this, it is surprising I was able to return to General Motors under such economic conditions.

In the Spring of 1933 we had a slack period in the design department and, instead of reducing the personnel, Harley Earle set up a design contest among the designers. He divided the small staff into teams of three men with one being named the chief designer for that group. In my case, he made my team four in number as two were apprentices. There were five teams. The leaders, besides myself, were Frank Hershey, Jack Morgan, Jules Agramonte and Tom Hibbard. My team included Carl Otto.

The rules for the contest specified a four-door sedan to be exhibited in one-quarter scale model form.

I had an inspiration for an unusual design based on my experience with the Model A Ford. I always tried to keep it in show condition, including the engine. Keeping the engine clean was a lot of work and I told my group how we could design a car in which the engine would stay clean. The scheme was to have two radiators located between the hood and front fenders and to block off direct air flow to the engine compartment. It wasn't a very good idea but it set up a package problem that was bound to give a new look to the front end.

There are no pictures of the finished model, but I recall the team and myself were happy with its appearance.

A prize was awarded to the winning team. It was a trip to Chicago for several days to attend the World's Fair going on there, called A Century of Progress Exposition. A generous amount of spending money was to be given each member of the team.

The judges of the contest represented some of the General Motors Divisions and a couple of the Fisher brothers were also judges.

In the official judging, my team took last place. Prior to this, however, they had an unofficial judging in which all of the contestants cast votes for their number one selection, not including their own creation. In this, we won first place. Jack Morgan's group went to Chicago and they did have a good design.

On September 20, 1933, I had a call from Harold Ames saying he had an idea he wanted to discuss and invited me to come to Indianapolis for that purpose over a weekend.

Ames had an idea how a new smaller and much lower priced Duesenberg could be produced while the Model J was also continued. He got the idea from General Motors.

At the time, General Motors had the LaSalle in production under the Cadillac Division banner, although it used substantially an Oldsmobile chassis. This fact was not advertised but was well known in the trade. The arrangement gave Cadillac a lower-priced model in the depression market. The chassis was not very exciting, but the Fleetwood coachwork was unique and had Cadillac quality, plus the advantage of the Cadillac sales force.

Ames realized a similar arrangement could be available to him, since Duesenberg and Auburn were both part of the Cord Corporation.

The Auburn straight eight was a good car and had its own fine reputation for speed. Ames suggested we could work out a Duesenberg chassis from this basic machine and, by adding a tricky body, produce a sensational small, fast vehicle worthy of the name Duesenberg. Yet, it could be priced low enough to succeed in the depression market. It was a very sound plan and I agreed to return to Duesenberg to do the design work.

I resigned from General Motors for the second time.

The project was to be done in complete secrecy. Herb Newport and Alex Tremulus were doing the body design work on the Model J Duesenbergs and I was not to be concerned with it, nor were they allowed to help on my work. Phil Derham was to be the body engineer on the new car. Augie Duesenberg was now working for the company and was in charge of the chassis work in a small area of the plant which was sealed off with the same secrecy precautions that were imposed on my studio office.

I told Ames about my idea of a hermetically sealed engine compartment to keep the engine clean. I explained how the twin radiator arrangement which had been turned down by General Motors would give a new look to the front end of the car.

Ames didn't think the clean engine idea would prove of interest to the automobile buyer, at least not enough to make a new model successful. However, he was intrigued by the fact it was a design whereby you didn't have to build around a radiator at the front end.

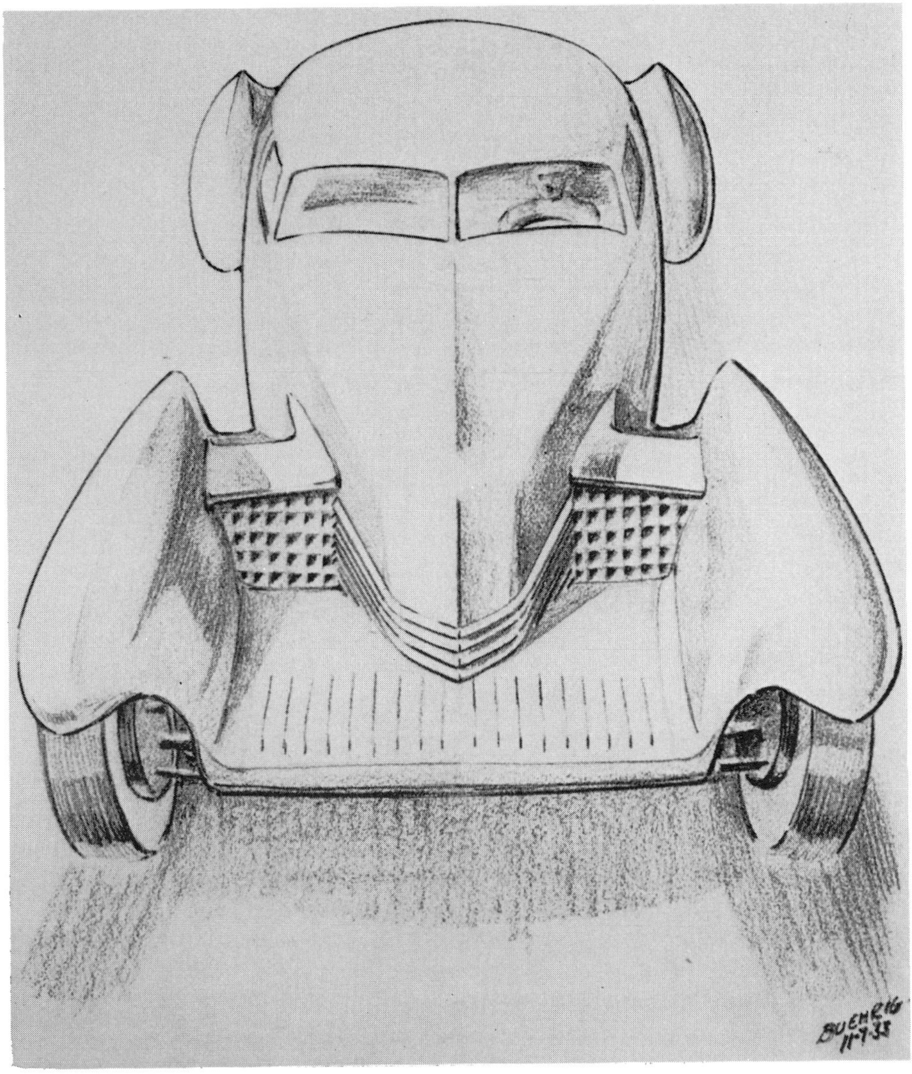

Above and below are the two pencil sketches Gordon Buehrig drew and showed to Harold Ames as his idea for a smaller and lower priced Duesenberg. The sketches are dated November 7, 1933. They show a vehicle with outrigger radiators which would allow a sealed engine compartment—a pet Buehrig idea. Refinement on these sketches resulted in the smaller Duesenberg prototype.

I made two pencil sketches for him and he approved the general idea. I worked out a one-eighth scale drawing of the design and then built it in clay in the same scale. Working in clay proved the ideal way for me to design. I was never very good at artwork either sketching or in rendering, but had a natural talent for sculpturing in clay.

Ames approved the clay model, an accurate orthographic drawing was made from it and turned over to Phil Derham.

The body was built by Weymann in Indianapolis and Derham followed the details of its construction. While this was going on, Augie Duesenberg was reworking an Auburn chassis and resolving the problems of the twin radiator arrangement. He used two small fans which were belt driven and the system worked but proved inadequate under high ambient temperatures.

The finished car was a real sensation but only a few of us ever saw it. This was in February 1934 and a set of circumstances were developing within the company and the automobile market which caused a change in plans. Had this not occurred, development of the small Duesenberg would have undoubtedly gone forward, the car marketed and have been successful.

These two views show the completed prototype for the small Duesenberg. It had an Auburn engine and rear drive. Radiators were kept outside as per the original sketch. The car was a sensation to all who saw it, but was shelved when Gordon Buehrig was pulled off the project to design the 1935 Auburn.

Another view of the small Duesenberg.

On May 17, 1934, Gordon Buehrig filed for a patent on his design of the smaller Duesenberg, and the patent was issued October 2, 1934. The design was, naturally enough, assigned to the Cord Corporation and later became the starting point for the Cord 810.

Oct. 2, 1934. G. M. BUEHRIG Des. 93,451

AUTOMOBILE

Filed May 17, 1934

Fig. 3

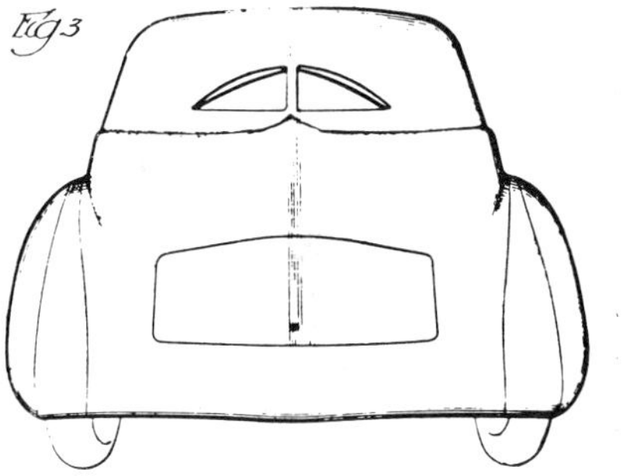

Fig. 2

Fig. 1

Inventor

By Atty.

Fig. 4

UNITED STATES PATENT OFFICE

93,451

DESIGN FOR AN AUTOMOBILE

Gordon Miller Buehrig, Auburn, Ind., assignor to Cord Corporation, Chicago, Ill., a corporation of Delaware

Application May 17, 1934, Serial No. 51,828

Term of patent 7 years

To all whom it may concern:

Be it known that I, Gordon Miller Buehrig, a citizen of the United States, residing at Auburn, in the county of DeKalb, and State of Indiana, have invented a new, original, and ornamental Design for an Automobile, of which the following is a specification, reference being had to the accompanying drawings, forming a part thereof:

Figure 1 is a rear perspective view showing the automobile of my new design;

Figures 2 and 3 are front and rear views respectively of the automobile; and

Figure 4 is a side view of the automobile.

I claim:

The ornamental design for an automobile substantially as shown.

Signed at Auburn, Ind., this 9th day of May, 1934.

GORDON MILLER BUEHRIG.

7 | Auburn

Auburn Automobile Company had spent half a million dollars in 1933 to design and engineer the 1934 Auburn line. Several executives within the Cord organization had been critical of the design, but they were ignored. The car was shown to the Auburn dealers in the winter and again at the January automobile show. It met with little enthusiasm and was a sales disaster.

Harold Ames had been one of those critical of the 1934 Auburn. He was called to Chicago early in 1934 and told if he could do something to improve the acceptance of the new Auburn, he had the job. Ames said yes, was given the title of Executive Vice President of Auburn and a budget of $50,000 for tooling to cover a crash project to redesign the car for better public acceptance.

Ames moved from Indianapolis to Auburn. Augie Duesenberg and I came with him. Augie went to work with the Lycoming Engine Division of Cord and Louis Schwitzer of Schwitzer-Cummins to come up with something that would give the "new" Auburn more of a performance image. The redesign job was mine.

Designer Al Leamy, who had designed the 1934 Auburns, the original Auburn boattail speedsters and the first Cord - the L-29.

Although the company spent nearly half a million dollars to produce the 1934 Auburn, the public rejected the cars and they did not sell. This is a factory promotional photograph of the convertible coupe — not an unattractive car.

Front view of a 1934 Auburn cabriolet, designed by Al Leamy, showing the grille treatment which did not prove popular.

The 1935 Auburn, shown here in cabriolet form, was essentially a facelift - accomplished by Gordon Buehrig and his staff for a total of $60,000.

The four-door sedan 1935 Auburn with the Buehrig-designed facelift. The more attractive front, slightly reshaped fenders and elimination of the traditional Auburn belt molding curving up over the hood are the most noticable changes.

With a $50,000 budget, we couldn't do much. The decision was made to do nothing to the chassis or body and to concentrate on the front end sheet metal and fenders. Both front and rear fenders were changed, with a deeper skirt on both. We designed a larger, more impressive radiator shell and radiator grill and a new hood on which we eliminated the longtime Auburn trademark of the body sidebelt running into the hood. There were also new louvered side panels to go with it. A new radiator ornament and some other minor detail changes rounded out the facelift.

By June, the production line changes had been made and the "new" Auburn, called the 1935 model, started to sell. The performance image was enhanced by the introduction of the supercharged straight-eight. Lycoming made the engine and Schweitzer provided the supercharger.

Thus, the Model 850 Auburn arrived.

The fact the face lifted 1935 Auburns had been a mid-year introduction was great but it didn't give us a new "bomb" for show time when

It was leftover bodies for these 1933 Auburn Speedsters—an Al Leamy design—which originally prompted the design of the 1935 Auburn 851 Speedster by Gordon Buehrig.

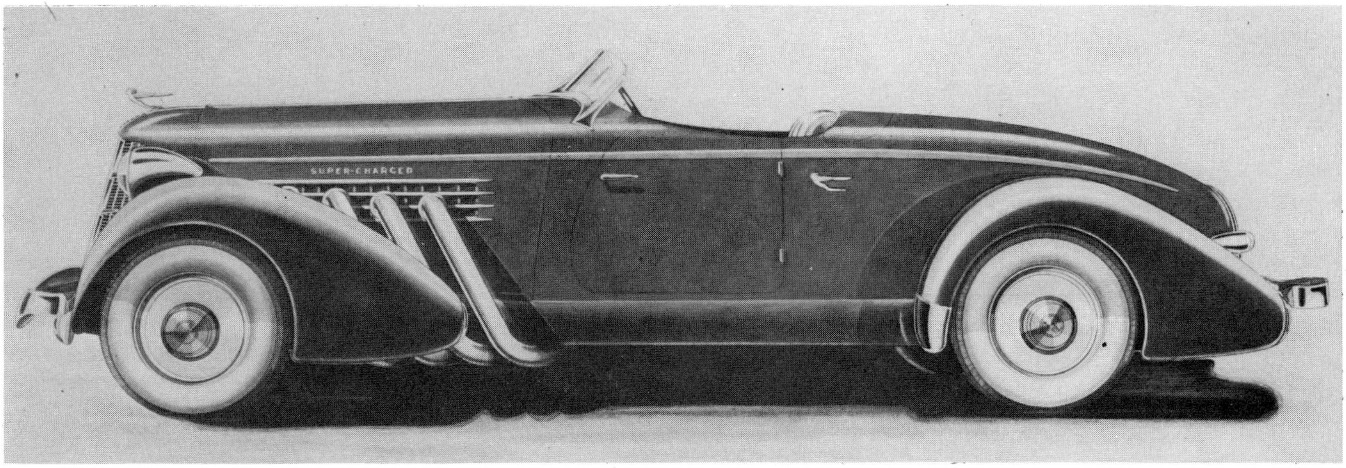

This early release on the new 1935 Auburn 851 Speedster was a photograph of the Paul Reuter-Lorenzen drawing of the car. The car was guaranteed to exceed 100 miles per hour with its supercharged Lycoming straight-eight engine. It is dated 12/20/34.

the other automobile companies were making their normal 1935 model line announcements. Ames wanted a sensational model to gain the free coverage of the media which you can get at Auto Show time if you have something worthy of this attention.

Auburn had built speedsters in the past. The last had been in 1933, an Al Leamy design. The bodies for these were built by the Union City Body Company of Union City, Indiana. Ames knew the body company had about a hundred of these bodies left over which had never been scrapped. His idea was to adapt a few of these to the 1935 Auburn chassis and display them at the Auto Shows. He asked me to see what I could do.

We had one of the bodies brought into our shop and determined since the body at the cowl was narrower on the 1933 model, we would need a special hood. The frame width was different and had a different shape at the rear axle kick up. This required compensating pieces to mount the 1933 body on the 1935 frame.

Ames had been pleased with the boattail speedsters I had done on the Duesenberg chassis, so I decided to make the Auburn similar in design.

The 1935 Auburn 851 Speedster in light finish is seen here in a foreign photograph.

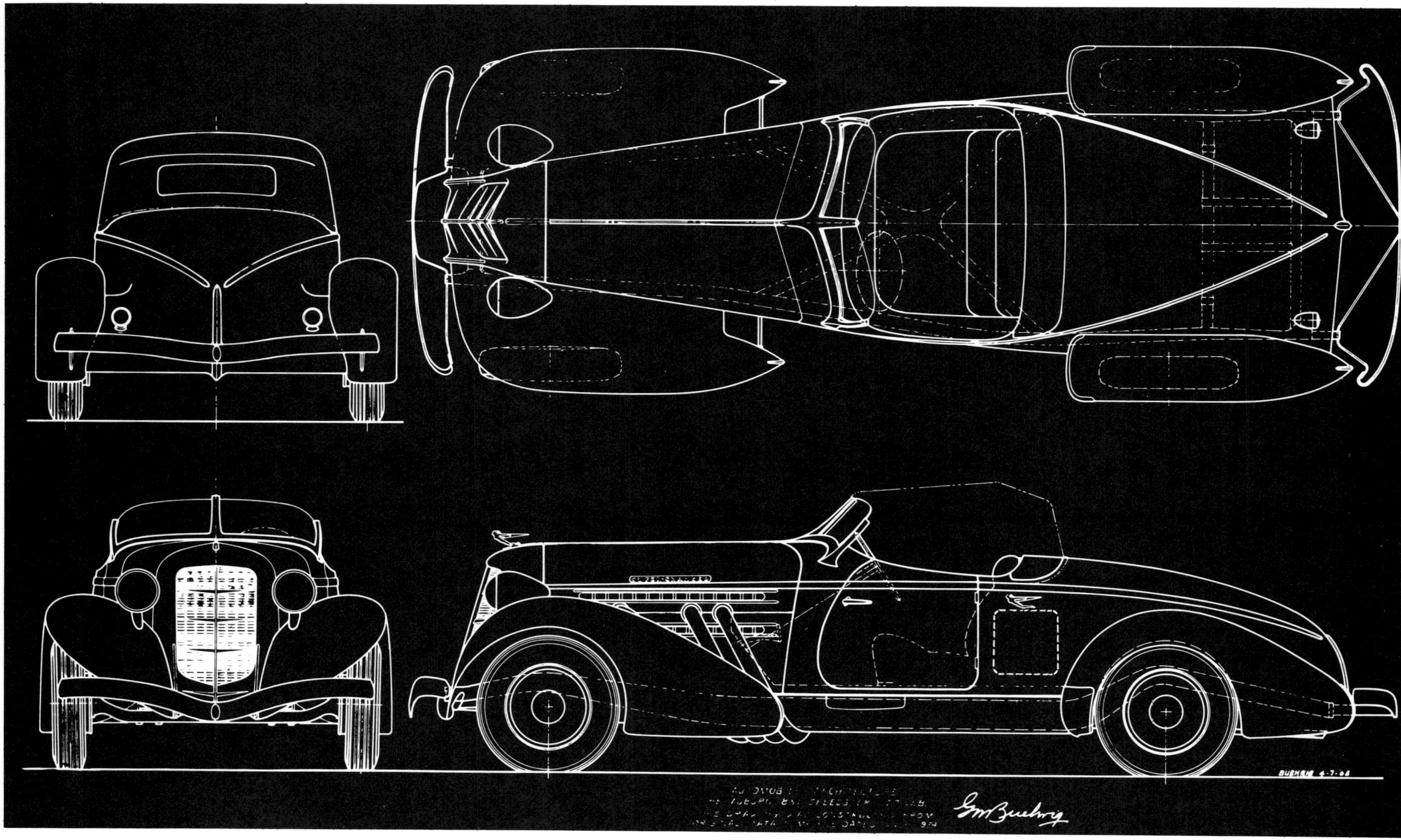

These drawings of the 1935 Auburn 851 Speedster were done by Gordon Buehrig in 1968 and based on information from his files, dated July 1934.

A most attractively restored 1935 Auburn 851 Speedster as shown recently at the Henry Ford Museum.

By using my group on a heavy overtime schedule, we did the entire design job in two weeks. We had the body cut in half just to the rear of the top well cover and threw away the rear portion. We built up a wooden armature to hold clay on the tail section and sculptured the new area in full size. We also sculptured the front and rear fenders on the mockup, setting it all up on a 1935 chassis. The new hood to match up the 1935 radiator shell and the 1933 cowl was added. The outside exhaust pipes completed the picture. Ames bought it.

Four were quickly built by hand and rushed to the four major Auto Shows. A price of $2,100 was high enough to show a good profit, even on the handmade models. This price sounds low today but at the time our lowest-priced sedan was only $595 and our supercharged four-door convertible phaeton was $1500.

The reaction to the Speedster at its introduction was exactly what Ames had anticipated. It stole the show and much of the newsprint. They sold several at the show and additional orders started coming in. From then on, they were built in small lots, ten or twenty at a time. They were always largely handmade with only very cheap hardwood draw forms used on a hydraulic arch press to form the body tail end panels and the fenders. The popularity of the model was much greater than anticipated by some members of management and when the inventory of 1933 speedster bodies was used up, they had to order more. About 600 were built when they decided to stop manufacturing them, although there was still some demand.

This picture of the instrument panel layout on the 1935 Auburn 851 Speedster shows not only instruments, but the plaque guaranteeing 100 mph and signed by race driver Ab Jenkins.

This high angle view of an early 1935 Auburn 851 Boattail Speedster shows the attractive overall lines to good advantage. Note the radiator ornament motif is repeated on the side of the body just above and to the rear of the door.

In the chapter on Duesenberg bodies, the difference in surface development of the rear end of the Duesenberg and the 1935 Auburn boattail speedsters is outlined. The design problem was substantially the same on both cars. In the case of the Duesenberg, the method used was the design of emperical lines on a orthographic drawing and the surface developed geometrically. In the case of the Auburn, the surface was sculptured in clay.

On the Auburn I got a better design because of the method used. Little design details like the chrome edge on the fender tips, the chrome stone shield on the front of the rear fenders, the split hood ornament on the body quarter panel and the chrome trim on the rocker panel on the Auburn were all copied from the Duesenberg. The windshield, doors, top and top mechanism were all from the 1933 speedster and the instrument panel was from the 1934 Auburn, all of which were designed by Al Leamy.

The Auburn Model 852 Supercharged Speedster in the 1936 model year was essentially a continuation of the 1935 Model 851 unchanged, except for the number 852 on the radiator grill!

Auburn enthusiasts say the 1936 Auburn 852 Supercharged cabriolet is one of the most attractive Auburn body styles outside of the Speedster.

Note this removable hardtop coupe version of the 1936 Auburn 654 line.

The 1936 Auburn 852 Supercharged line was easily distinguishable from the non-supercharged version, as it shared the outside exhaust pipes of the speedster and carried a "Supercharged" logo on the side of the hood.

The new radiator ornament, which was designed by a man in Buehrig's department, was part of Gordon Buehrig's facelift for the 1935 Auburn, as seen above. The split ornament on the body quarter panel was copied from the Duesenberg Speedster he designed and was carried out on both the Model 851 and 852 Speedsters.

In 1934, at Auburn, Gordon Buehrig designed this small rear-engined car as a possible future small Auburn.

A 1935 Auburn 851 Supercharged Speedster is seen here in the movie "Just My Luck", with Ann Grey and Charles Ray.

A 1936 Auburn 852 Supercharged Speedster, just sold to Indian Prince of Hyderabad, left. With him is E.Z. Ladovich, Auburn-Cord-Duesenberg dealer in Paris, France.

Actress Mary Astor and actor George Murphy (later U.S. Senator Murphy) pose here with a pair of 1935 Auburn 851 Supercharged Speedsters. Auburn frequently used actors for publicity photographs.

Cowboy star Tom Mix poses with a 1935 Auburn 851 Supercharged Phaeton.

Designing the Cord 810-812 | 8

After the emergency of the 1935 Auburn face lift program was out of the design department a decision was made, presumably in the Cord Corporation's Chicago headquarters, to revive the small Duesenberg program but to do it in Auburn, to redesign it for a V-8 engine and front wheel drive configuration and to call it a Cord.

The opportunity to work out the design of the new Cord and to have it a front wheel drive vehicle gave me an assignment as ideal as an automotive designer could imagine.

The drive line configuration on the new front drive unit produced a near-perfect package layout from a body designer's standpoint, as well as overcoming the mechanical shortcomings of the L-29 Cord layout.

On a front wheel drive car the centerline of the differential is usually within a fraction of an inch of the centerline of the front axle. It is kept slightly off center to keep the bearings active at all times.

In the L-29 Cord layout, the driveline configuration saw the differential substantially in line with the centerline of the front axle.

The 1930 Cord L-29 Convertible Coupe; the first front-drive Cord.

Behind this came the transmission, followed by the clutch and flywheel and then an eight-cylinder in-line engine. In this case, this arrangement put the dash (the front of the body) approximately 55 inches back of the centerline of the front wheels. Putting the engine so far back moves the center of gravity of the vehicle back as well. The result on the L-29 was the driving wheels had a severe tendency to spin on acceleration on gravel or a slick road surface.

When a vehicle is driven up a hill the center of gravity, which is above ground level, tends to move even further back and this under conditions where maximum traction is desired for the driving wheels. L-29 Cord owners were aware of this but most of them loved their car so much for its sheer beauty they forgave its faults.

On the proposed new design the transmission was forward of the differential and the clutch. The flywheel and engine were behind. The engine was a V-8 which is about half as long as a straight eight. This configuration was accomplished by locating the centerline of the crankshaft above the centerline of the differential. The drive to the transmission was located at a higher level and returned from the transmission to the differential at the centerline of the front axle. This layout resulted in moving the center of gravity forward and reduced the space from the front axle to the dash by approximately 22 inches as compared with the earlier L-29 Cord. In short, it provided a near-perfect chassis layout as far as my work was concerned.

Roy Anderson, chief body designer, Ted Allen, chief body draftsman, and Bart Cotter, assistant chief body draftsman, all wanted to engineer the car as a unit body frame construction. This type of

A factory photographed 1930 Cord L-29 with standard four-door sedan body. These attractive cars were the design work of Al Leamy.

In the L-29 Cord layout, the driveline configuration saw the differential substantially in line with the centerline of the front axle. It is the same layout that race car builder, Harry Miller had used in his front drive cars.

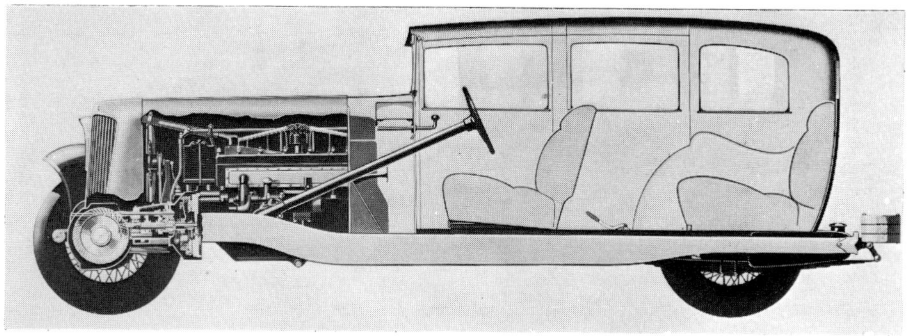

Front end of the chassis of the L-29 Cord, showing the engine, transmission, differential, front axle, front brake drums, spring suspension and steering universal joints.

construction gave me an optimum body layout for my work. It enabled me to use a step-down floor design which was a feature "invented" with much fanfare by Hudson 12 years later.

When the new Cord assignment was given to engineering the excitement was contagious. My group simply forgot the clock. Overtime pay was unheard of ,yet we worked several nights a week. We had a radio and enjoyed the Fred Allen shows, Amos and Andy and other favorites as we worked. There simply wasn't anything we could do that would have been as much fun as working on the Cord.

I got beautiful cooperation from the chassis department in the location of such items as the battery, the muffler and other components I wanted out of the way so we could maintain an uncluttered floor level.

Cooperation was also perfect with body engineering because I kenw and understood their problems. An illustration of how this two-way-street worked involved the doors. One day, Harold Molliter, a body draftsman, asked if he could show me a scheme he had developed which would allow the stamping of the door outer panels without the use of cam dies around the window openings. In order to use it the sheet metal had to be formed in a very specific way which was different from any other car on the road. He wondered if I would like the effect. We tried it, liked it, and used it thereby saving a substantial amount of the cost of the door tooling program. We also later discovered this door design detail had an additional virtue of cutting down wind noise.

The body was all steel construction. Normally, in a body program about half of the total tooling cost is for the doors. On our program this

A picture of the Auburn Auto Design Department, taken in September 1935, showing, from left: Paul Peter Rueter Von Lorenzen, airbrush artist; Art Kruger of sales, used here to fill in the photo; Dick Robinson, working on the Gentlemen's Speedster; Gordon Buehrig; Vince Gardner at the desk, and Dale Cosper, working on the phaeton model of the Cord 810. Note the styling bridge and model building equipment on the center table. It was invented here.

was cut way down because of not using cam dies and because I laid out the body in such a way that the left rear door was made from the same tooling (plus one extra operation) as the right front door. The same was of course true of the right rear door sharing tooling with the left front.

The method we used in developing the design was unique. It involved invention and was probably the quickest and least expensive production car program ever accomplished in an all-steel body.

Dick Robinson laid out the complete package and chassis engineering requirements for me on a one-quarter scale drawing. On this orthographic drawing I drew in all the emperical design lines, just as I had done when doing body designs at Duesenberg. When this was completed a print of it was given to Vince Gardner who designed a wooden clay buck armature with a minimum of one inch left on all surfaces for clay.

The type of clay used in automobile design work is a temperature sensitive material which at about 110° is soft and pliable, but at normal room temperature is very firm and can be worked with extreme accuracy with steel tools. In using it a quantity of material is kept at 110° Fahrenheit in an electric oven so it is always available to the sculptor.

On another working print from the one-quarter scale layout I spotted all the points on the car where minimum size requirements had to be held or exceeded. There were dozens of such points, including head room requirements above the front and rear seats, air cleaner clearance under the hood, radiator clearance, the extreme travel of the wheels in full jounce, etc. Each of these points were located on the drawing in each of the three planes of projection, i.e. side elevation, plan view and end elevation.

Our "invention" while working on the Cord design project was the styling "Bridge." The bridge is now a common tool used every day in all automobile studios on both full size and scale model design work. I may be wrong, but in all the investigating over the years, it appears we did it first.

The bridge is an inverted U-shaped tool that travels transversely over the length of the car on guides or tracks. With it, any point on the surface

Dick Robinson, left, and Dale Cosper work on a model for one of the compromise programs. The original Cord model is in the left foreground.

of the vehicle can be instantly located in space. The bridge is marked with ordinance lines and so is the table on which it travels. To locate any point in space on a model, the bridge is moved so its face is square with the point. This gives the longitudinal dimension. The height of the point projected to the bridge gives that reading and the distance from the point to the vertical leg of the bridge gives the location of the point with relation to the longitudinal centerline of the car. If the bridge is (to scale) 80 inches wide or 40 inches from the centerline and the point is 10 inches inboard from the bridge vertical member, the point is 30 inches from the centerline. (40-10).

The use of the bridge is described in detail because it was a key to the speed with which we completed the design program. The equipment was made to our drawings in the pattern shop.

The clay buck was properly located on the design table and then we all pitched in claying it up. We had bandsawed a few rough contour templates for guides and we purposely overclayed it so there would be excess material to remove.

The next step was to establish all of the critical minimal dimension points which I had established on the working drawing. We did this with 1/8 inch dowl sticks 1/2 inch long. Using the bridge for location and referring to the drawing, we drove these dowls into the clay so their top point exactly marked the critical dimension. Each minimum critical point was so marked. Now our engineering space problem was spelled out for us on the model. I could shape up the surface and know I was not shy on any dimension as long as I didn't go below any of the dowl points.

When I had one side of the model sculptured, we used the bridge and its accessories to duplicate the design on the other side.

The outside face of the bridge was used as a plane of reference and measurements from it to different points on the finished surface were located on the unfinished side by this method. Parallel 1/4 inch holes were located through the bridge uprights and one and one-quarter inch intervals and through these we would insert a probe on the finished side of the model so its tip end touched the surface. A locating collar with a quick-lock adjustment on the probe on the outside of the bridge was locked in place. Then the probe was removed and the tip, which was turned down to 1/8 inch diameter with a rough filed face, was marked with blue carpenter's chalk and then inserted through the corresponding hole on the unfinished side of the model and pushed into the clay until the collar came against the bridge. This process was repeated over the entire model every one and one-quarter inches. The result was the clay on the unfinished side was marked every one and one-quarter inches to the proper surface. The clay on the unfinished side was then scraped away until all the blue dots appeared on the surface. This process gave us a fidelity of surface to 1/32 inch accuracy.

The one and one-quarter inch dimension on a one-quarter scale model corresponds to five inches on the full-sized car. When the model was completed and approved, the bridge was again used in a similar way on one side of the model to give a reading in inches from each point on the surface and by using a one-quarter scale, the dimensions were taken in full size.

The ordinance points were recorded in chart form and, when completed, it described for the full-sized body draft the entire surface of the vehicle.

In order to save time and money, we went from the one-quarter scale design model direct, through the method just described, to the full-sized body draft and die models. Bart Cotter did most of the body surface drafting and Ted Allen and his draftsmen did the structure and detail drafting.

I designed the interior in one-quarter scale. This included the instrument panel, which gave me the opportunity to do one of my favorite parts of a car exactly as I had always wanted to.

I used gauges of different sizes and of different shapes but always limited to production units except for the faces. This arrangement made it possible to identify any specific gauge much easier than if they had all been round and of a common size.

Gordon Buehrig designed what he called a "no nonsense" instrument panel for the Cord 810, set off in an engine turned panel. The circular panel mounted above the windshield is the radio speaker, which Gordon Buehrig says was inspired by the French Marchal headlight design.

It was a "no nonsense" panel, made strictly to give the driver the information he wanted at a glance and it was located high on the cowl for best visibility.

The Magnavox Company in Ft. Wayne, Indiana, had a new (at that time) method of illumination of gauges called edge lighting. A colored dye was applied to the edge of the glass over the gauge and the lighting so located and shielded that all light to the gauge would go through the dyed edge. I spent several hours in their darkroom laboratory selecting the exact color. I also equipped the instrument panel light switch with a rheostat so any degree of light intensity the driver wished was available. This also was an automotive first.

A fan-shaped gauge with an energizer button switch gave a direct reading of the oil level in the crankcase, which eliminated having to raise the hood to check the oil.

I used the old King Seeley hydrostatic tube gauge which looks like a thermometer to show water temperature as reading temperature on this type gauge is such a common and therefore natural reaction.

The tachometer added worthwhile interest to driving a Cord. The car had a four-speed transmission and the tachometer told the driver, for instance, that at 65 miles per hour, driving in fourth gear, the engine was only turning 2,000 revolutions per minute and there was a lot more power available if he cared to use it. Third gear was direct and was used for city driving, while fourth gear was an overdrive gear and used for highway cruising. All these things added to the excitement when driving a Cord. As soon as you started the engine all the gauges came to life and you knew you were in command of a fine piece of machinery that wanted to GO and to GO FAST if you cared to.

The light switches and the manual choke were control levers in a case in the center of the panel, inspired by the hand throttles of a multi-engine airplane.

The gearshift lever was not on the instrument panel but it was so close to it that it was part of the same picture. The lever itself was only about two inches long and was housed in a gate case on the end of the housing attached to the steering column. The placement of the lever allowed shifting gears with the third finger of the right hand without taking your hand from the steering wheel. It was called the Bendix Electric Hand. In operation, it actuated electric solenoids which in turn operated vacuum cylinders which shifted the gears in the transmission. It functioned when the regular clutch pedal was operated and this made possible preselection of gear changing if the driver desired.

The Lycoming 90-degree V-8 engine, designed by Forest S. (Bill) Baster, chief engineer at Lycoming. Its 3 1/2 inch bore and 3 3/4 inch stroke gave 288.64 cubic inches displacement. The Bendix transmission shifting servo is on top of the gear box.

This invention by Bendix was a timely coincidence in the birth of the 810 Cord. It solved what would have otherwise been a knotty problem.

The engine and driveline configuration which put the transmission in front of the radiator made it so far from the driver and so inaccessible that any type of mechanical linkage control would have been difficult. The Bendix Electric Hand was an ideal solution because with it, only electric wires were routed around the radiator.

The mechanical parts of the Bendix unit which were mounted on top of the transmission and forward of the radiator created a styling problem because they projected above the normal sheet metal in this area. On the experimental Small Duesenberg this problem didn't exist because of the conventional rear drive layout and the result was just a plain and very uninteresting area of sheet metal between the front fenders.

On the Cord, I took a very direct approach and allowed the problem to solve itself. Engineering had specified a one inch clearance between the mechanical parts of the Bendix unit and the enclosing sheet metal. It was an asymetrical layout and we built up a wooden mockup of the problem area showing the side of maximum interference. Over this we laid on extruded strips of hot clay one inch thick and then smoothed up the clay, transferred the same design to the other side making it symmetrical and the esthetic effect was excellent.

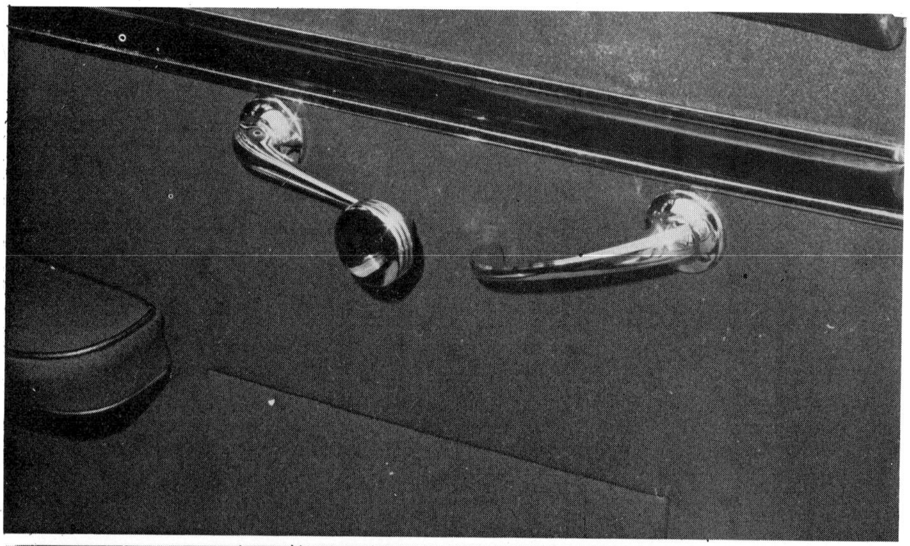

Cord 810 interior hardware was again a money-saver. Gordon Buehrig selected an already-existing set of handles from Doehler Jarvis Company, then made up new plastic knobs from a new Eastman plastic called Tenite.

The Cord 810 steering wheel, which was lauded as an innovation was, in fact, a stock wheel which could be purchased without tooling cost, plus a Buehrig-designed horn ring and button.

Steering wheel molds at that time cost about $10,000 and I knew we couldn't afford one. I went to the Sheller Manufacturing Company in Portland, Indiana, and looked at all of the obsolete steering wheels on which Sheller still had the tooling and selected a very simple design. Back in Auburn, I designed a horn ring layout for it and a special center hub cover which, when added to the old Sheller wheel, created a sensational new steering wheel and started a styling trend in the industry which lasted many years. Mercedes had been using a horn ring for several years before the Cord, but it was a new look for an American car.

A similar situation existed in the problem of the interior body hardware. Interior door lock handles and glass regulator handles are made in die cast molds and the tooling is expensive. This time I went to Toledo, Ohio, to the Doehler Jarvis Company and spent a day looking at all the obsolete hardware they had which we could buy without a tooling expenditure. I selected a design which was not too bad and back in Auburn I designed a large one and three-quarter inch diameter colored knob for the window regulator. The knob was color keyed to the interior trim scheme and was so much in evidence that the Cord appeared to have a new design of hardware. Poly acetate butyrate, one of the first modern thermoplastic materials, had just been introduced by Eastman with the trade name "Tenite." It was a great new material (and still is) with an excellent color range. It was also an easy material to machine and therefore ideal for making sample parts.

John Slater, the Eastman salesman, provided us with a bar of Tenite two inches in diameter and we turned out the knobs for the window regulators on our small shop lathe. At some point in the production of the Cord these regulator knobs were die cast and painted. It was cheaper than Tenite and the effect was the same, at least until the paint wore off. Fred Duesenberg used to say one trouble with American cars was that too much engineering was done in the purchasing department!

For the general interior trim design I copied the interior I had created for the Duesenberg Beverly a few years before.

In covering the transmission with a panel, we let "form follow function" and simply built the panel around the components, with the required clearance as specified by the engineers.

In designing the exterior of the car in rear elevation, I wanted to create a character which would be easily recognizable at night. I knew the new shape of the Cord solved this problem in daylight, but I had to do something with lights to solve this problem after dark.

The impact of some innovations on the Cord are lost to people today who do not remember (or were not even born at the time) what competitive cars were like. The standard practice at the time in taillights was a single bulb mounted on a bracket attached to the left rear fender. It had a red lens behind the bulb for a taillight and a clear lens below to illuminate the licence plate. On the Cord I mounted the licence plate in the center of the trunk lid and illuminated it with a bulb in a cantilevered housing above the plate. For rear lights I used two four and three-quarter inch diameter red lights, located on either side at the bottom of the trunk lid. When you came upon a Cord at night, you knew it. Within two years this type of rear end illumination became standard on most cars.

License plate light and tail lights add a great deal of distinction to the back of the Cord 810.

Probably the most talked about innovation on the Cord was the disappearing headlights. We had used them originally on the experimental Duesenberg prototype. Ames was particularly proud of this innovation because it was an idea taken from the landing lights on the Stinson airplane. Stinson Aircraft was then a part of the Cord Corporation.

On the experimental Duesenberg, on the original Cord design and the first prototype, the lights folded out from the inner face of the front fenders. Before the car was released for production they were relocated so they folded out from the center of the fender, which was a better location. We researched the idea of having them operate by electric motors but the design and development cost was too high and we settled for operation controls by the use of two small cranks at the extreme ends of the instrument panel.

Just about the most noted thing about the new Cord was the pop up head lights. Many cars have had the disappearing headlights but Cord had them first.

The supercharged Auburn Speedster.

1936 COF

O Sedan

1937 Cord 812 Supercharged Phaeton.

On the Cord, we pioneered the use of large hubcaps which covered the entire wheel. Prior to that some discs had been sold as accessories to cover wire wheels because of the trouble keeping wire wheels clean, but we were the first with full wheel covers for steel wheels and to have them as standard equipment. On the first design the shape of the wheel covers was the same as that which went into production except the twelve one and one-half inch diameter holes around the perimeter of the covers were not there. In testing the prototype vehicle, a problem of brake failure was attributed to a lack of ventilation of the brake drums. To prove this, in our experimental shop, they put in the holes and solved the problem. They explained this to me to see what could be done. I saw the altered wheelcovers, liked their appearance and they were released for production. Once again, as LaCorbusier says, form follows function!

Radios were becoming a popular accessory for automobiles and we decided to make them standard equipment on the Cord. I located the radio speaker in the roof at the top of the windshield on the sedan because in this position it projected sound to the rear compartment as well as to the front. For the bezel of the speaker, I worked out a design clearly inspired by the Marchal headlamps then so popular in Europe. The radio antenna was located beneath the car so the exterior design would not be cluttered up with a buggy whip aerial. You may wonder why this location for the antenna is not in use today. The answer is the buggy whip type aerial works better. The radio controls added one more functional bit of detail to an already impressive instrument panel.

On the Cord 810, another new feature was one of the first uses of large hub caps which covered the entire wheel. The round holes in the cap were dictated when testing showed caps without holes caused brakes to retain heat.

Marchal head lamp design inspired this Cord radio speaker.

The front and rear bumpers on the car were somewhat different and were designed by the bumper manufacturer, Buckeye Bumpers of Springfield, Ohio. Augie Duesenberg was working at Auburn on the problems of producing a supercharged version of the Auburn. The vendor from Buckeye Bumper showed Augie a sample bumper which we could buy without tooling cost. We put it on a prototype car, liked it and approved it for production.

One other small item which was an automotive first was the sheet metal door covering the gasoline filler cap. It was equipped with a lock since stealing gasoline was not uncommon during the depression.

Graceful yet functional bumper of the 810.

Cord 810 Patent Application Drawings and data.

UNITED STATES PATENT OFFICE

97,697

DESIGN FOR AN AUTOMOBILE

Gordon Miller Buehrig, Auburn, Ind., assignor, by mesne assignments, to Auburn Automobile Company, Auburn, Ind., a corporation of Indiana

Application August 5, 1935, Serial No. 57,946

Term of patent 7 years

To all whom it may concern:

Be it known that I, Gordon Miller Buehrig, a citizen of the United States, residing at Auburn, in the county of De Kalb and State of Indiana, have invented a new, original, and ornamental Design for an Automobile, of which the following is a specification, reference being had to the accompanying drawings, forming a part thereof.

Fig. 1 of the drawings is a top plan view of an automobile, showing my new design and illustrating the oval-shaped casing elements of the collapsible headlights flush with the fenders, that is, in the position in which they are disposed when the headlights are in their inoperative or collapsed position. Fig. 2 is a side elevation. Fig. 3 is a front elevation. Fig. 4 is a rear elevation. Fig. 5 is a perspective of a portion of the front end of the automobile.

I claim:

The ornamental design for an automobile, substantially as shown and described.

GORDON MILLER BUEHRIG.

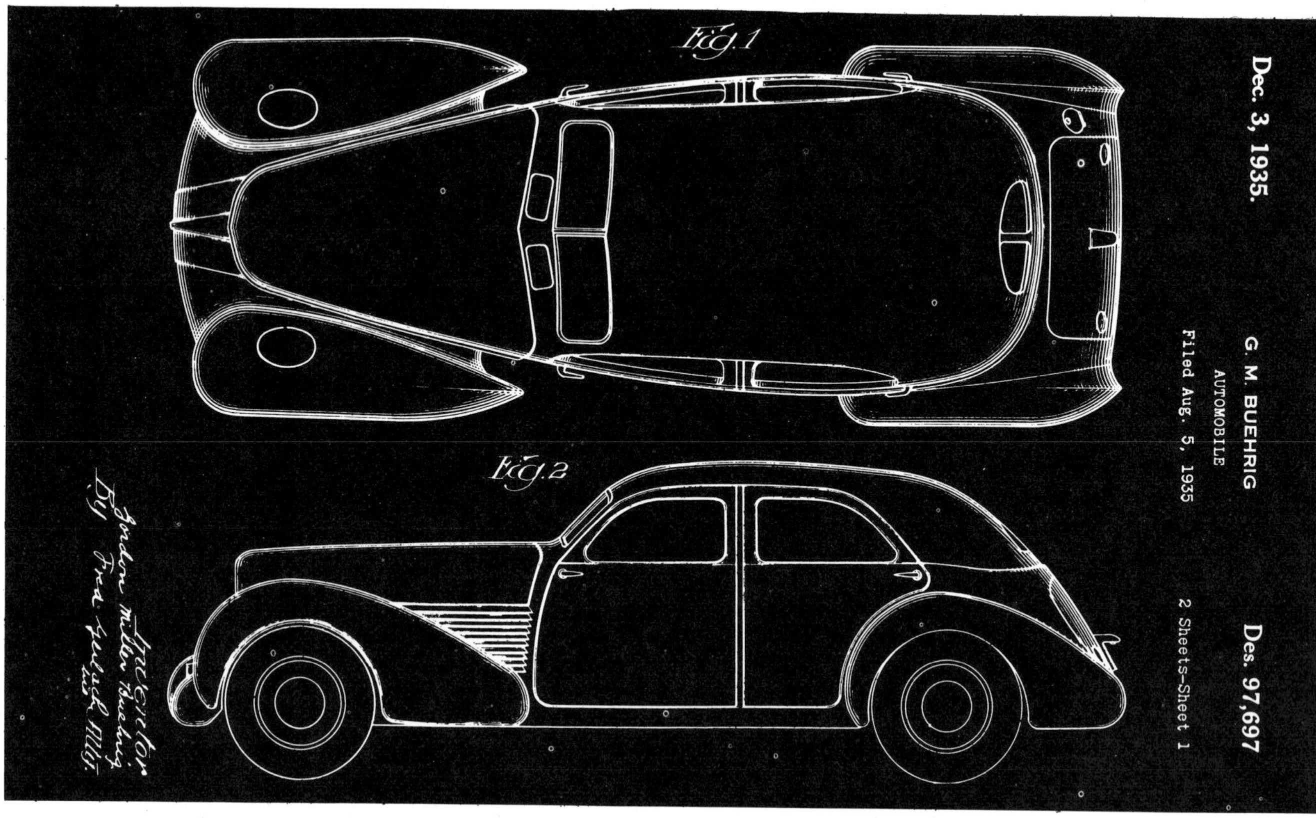

The 810 Cord transmission.

The interchangeable door design and the type of construction used for them had an additional virtue of cutting down wind noise.

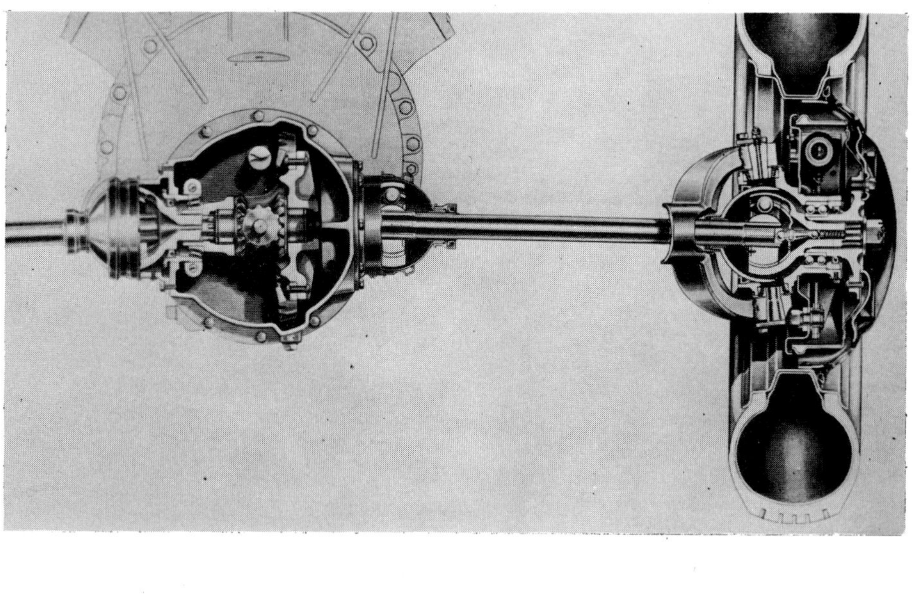

The differential and front wheel drive mechanism.

9 The Cord program: Death and resurrection

In my little diary book, the date December 22, 1934 carries a brief but bold statement. It says "Married Betty Whitten."

The year 1934 had been a good one for me. The 1935 Auburn facelift had been completed and had proven successful. The 851 Auburn Speedster project was also successfully completed. The design of the new Cord 810 was virtually completed by the end of the year and it was time for a breather.

I met Betty on a blind date the day after Labor Day 1934 and on our second date asked her to marry me.

After the ceremony on December 22nd, we left in our Auburn four door convertible sedan and drove to Florida. Everything was beautiful. The Cord project was completed as far as design was concerned. Basic engineering was finished and the die models for all the body stampings were completed.

When we left Auburn for Florida, it was with complete confidence that the Cord project was on schedule, running smoothly and, with this attitude, I forgot the automobile industry for the next two weeks.

Betty and I returned to Auburn from our honeymoon on January 6, 1935 and when I went back to work I learned the Cord program had been killed.

I can't remember all the details but do recall the question of whether the company had enough money to do the Cord program was a legitimate one and there was not complete agreement among company officials on what should be done. This resulted in a number of compromise programs we were asked to do and which we begrudgingly carried out. One such program was to use the Cord front end on the old

One of the famous Dale Cosper sepia prints of the Cord 810 design model which was shown to the Cord Corporation Board of Directors and helped make the decision to produce the car.

Dale Cosper photo of the rear view of the Cord 810 scale model.

A 3/4 front view of the Cord 810 scale model in one of the Dale Cosper sepia photos used for the presentation to the Cord Corporation Board of Directors.

A front view picture of the Cord 810 design model.

Auburn body. We in Design knew what a monstrosity this would be, but we had to build a one-quarter scale model to prove it.

After months of frustration while we worked on such ridiculous programs, something significant happened. It was on July 7, 1935, a Sunday morning. I received a phone call from Roy Faulkner. He told me there was to be a meeting the next day at Cord Corporation headquarters in Chicago. At that time, the Board of Directors would decide Auburn's future plans. He wanted to know if I had any good pictures of our Cord design model. I told him we didn't because I had prohibited any photographs of the model being taken to keep the project entirely confidential. I also told him I would see what could be done.

I immediately called Dale Cosper who, among his varied talents, was an expert amateur photographer. I picked up Dale and his little German Steinheil camera, at his home and we drove out to the plant. From noon until two a.m. Monday morning we did a photographic job that made automotive history. The Board of Directors decided to go ahead with the project and Faulkner told me later that Dale Cosper's pictures did it.

The date for the New York Auto Show was November 2, 1935. This left three months and 26 days to pick up the Cord program where it had been stopped the first of January and to build and test a prototype, complete tooling and have production cars ready for the show. It was an impossible task and although it was accomplished (after a fashion) it was not done well and the results were so financially crippling that it eventually put the company out of business.

The Automobile Manufacturers Association which sponsors the New York and other major auto shows had a ruling then that no automobile could be exhibited unless a minimum of 100 cars had been produced before the show date. Production tooling could not be made quickly enough to be used on the 100 cars so they all had to be made by hand. The power hammer department at Auburn worked as many hours overtime as the men could stand. Our vendors were asked to work miracles.

A program, rushed as this one was, is much more expensive than one with normal timing. Also, the Cord Corporation's financial position was insecure, which was no secret in automotive circles. The only way the program could work was to have Auburn's major vendors extend credit on the new commitments. For this purpose a party was held at the administration building in the showroom. At this party the prototype of the Cord 810 was unveiled for the vendors and this established sufficient faith in the project for them to meet Auburn's credit request.

The transmission for the new Cord was an entirely new design requiring new tooling. The time was too short to produce the tools and build the transmissions for the 100 show cars and consequently,

Top views of the prototype Cord 810 Sedan.

although the cars were built, they wouldn't run because they were without transmissions.

We saved a little time on the phaetons for the shows because they were designed with a disappearing top. This was convenient. It allowed us to exhibit the cars in the show with the tops down . . . and worry about building the tops later!

During these months of uncertainty and frustration, the 100 cars somehow were built and our exhibit was on the train for New York in time for the opening of the show.

To say the Cord stole the show would be an understatement. It would be more accurate to say the Cord was the show!

In an article by George A. Brewster titled "Body Design at the New York Show," dated December 1935, he said "The most unusual group of cars in the show—infact, in a good many shows—was the half-dozen cars comprising the Cord exhibit. In one or two respects this entirely new car is of greater interest than all other exhibits put together.

"In the matter of color and finish, the Cord people have given it more thought and study than other manufacturers. The general idea is to have the exterior of the car finished in a single color, with the interior of a properly contrasting hue . . . The results, from an artistic standpoint, are breathtaking.

"The entire exhibit of Cord cars provided one of the most aesthetically satisfying of the show."

Another article, titled "The Body Beautiful," stated "Cord has taken full advantage of its front-wheel drive in developing a decidedly unconventional exterior. Without a shaft or differential to interfere, it has been possible to lower the body until, despite its 125 inch wheel base, it stands only five feet high, and can be stepped into without the benefit of running board. Retractable headlights, the most unusual feature, give increased speed by reducing windage."

Another said "The designer has been flooded with congratulatory letters and cables from design engineers all over the world. The compliment the designer liked best, 'It looks as if it had been born and

A 1936 Cord 810 Sportsman photographed at the factory. It was probably one of the first. To get some idea how radical the Cord 810 design was at the time, compare it with the cars lined up down the street behind it.

A 1936 Cord Cabriolet Sportsman—the rumble seat version—photographed here at the factory.

A 1936 Cord 810 Coupe, one of only two or three built, is seen at the factory delivery point.

Interior of the 1936 Cord 810 Coupe, photographed at the factory. Note the roll-up curtain covering behind-the-seat storage.

raised on the highway.' And properly. For sheer taste, for functional correctness and for beauty the 810 Cord is the best design the American industry has ever produced."

At the New York Show the cars were on the floor rather than on raised platforms and because of the low height of the Cord it was easy to see where it was because of the crowd around it. But, it was almost impossible to get through to see the car. People were standing on cars in adjoining exhibits to get high enough to look over the crowd for a glimpse of the Cord. One sales person at the adjoining Chrysler exhibit was furious about people standing on the bumpers of the Chryslers to try to see the Cord.

At the National Auto Shows throughout the country, a number of Cord orders were taken for Christmas delivery. Why such promises were made is hard to understand but optimism was high . . . even beyond reason.

When the time came to inform customers their promised delivery dates could not be met, Harold Ames came up with a device that kept a number of orders on the books, although there were some cancellations. Ames had a firm in Buffalo which manufactured jewelry and was especially clever in the art of small bronze castings make a number of models of the Cord sedan in 1/32 scale. These beautiful models, which cost the company $20 each, were given to customers who had Cords on order before Christmas. Today these models are collector's items and trade at very high prices.

Production of the Cord got underway slowly and the first cars were plagued with trouble. The cars overheated in warm summer weather and the transmissions didn't work well.

During the early days of production, Cord's competitors took advantage of the problems with the car and spread rumors. It is surprising the car lasted two years. It did so simply because it was so well liked that people bought it even though they may have felt it was an unwise investment. Actually, the problems with the car were pretty well solved by 1937 and they were good cars by the time Cord went out of business. In recent years the restorers who have rebuilt the old Cords have solved all the problems which should have been corrected in the original development period. Some cars show up every year at the Auburn-Cord-Duesenberg Club homecoming in Auburn which are better automobiles than when they were new.

One of the first 1937 Cord 812 unsupercharged Beverly Sedans with the trunk added at the rear.

A 1936 Cord 810 Phaeton photographed upon completion at the factory. Note the unusual rack on the rear bumper.

A 1937 Cord 812 Supercharged Phaeton.

A rare 1937 Cord 812 Phaeton with dual sidemounted spares.

Factory photo of the 1937 Cord 812 Custom Beverly Sedan. This version, on the 132-inch wheelbase, had a division window between back and front seats.

The first supercharged Cord sedan was based on a standard Beverly.

Gordon Buehrig's personal 1937 Cord 812 Supercharged Phaeton, photographed in 1939.

A young movie actor by the name of Tyrone Power polishes his new Cord 810 Phaeton's instrument panel.

New Cord 810's were on hand for the 1936 Indianapolis 500, being used by various AAA officials.

Long-time Auburn and Cord record maker Ab Jenkins is photographed here at the factory in 1936 with a new Cord 810 Sedan.

Rear view of the Gentelmen's Roadster clay model, as shown at the New York Auto Salon.

The Gentlemen's Roadster model, one of Gordon Buehrig's design exercises. Harold Ames had suggested it as a possible vehicle for using up a number of left over Auburn V-12 engines.

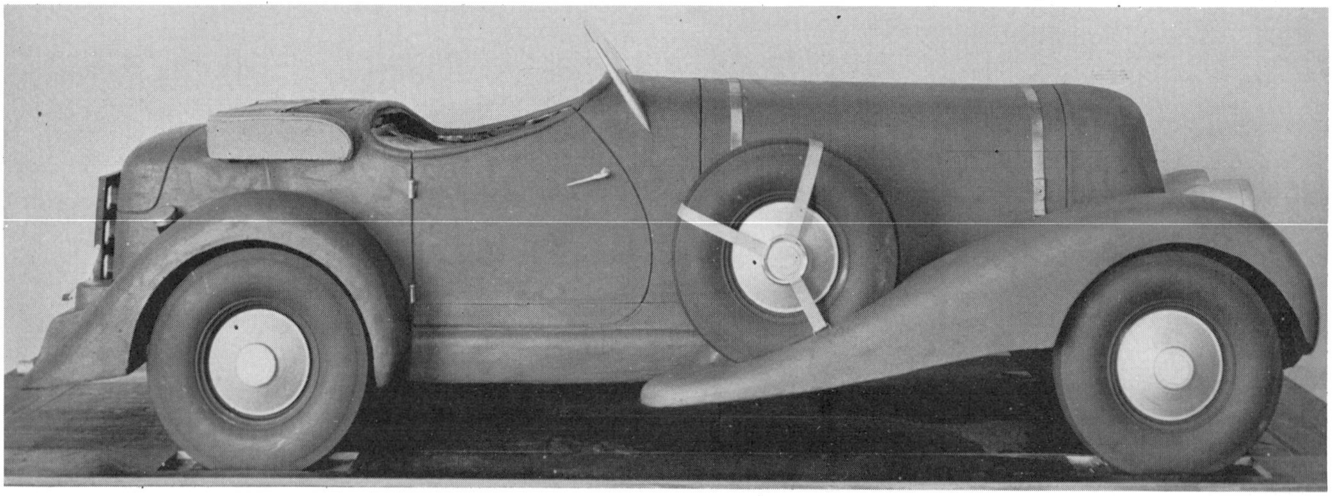

Auburn eventually built a prototype of the Gentlemen's Roadster, using the Auburn six chassis, with the V-12 set well back. It used a standard Model J Duesenberg hood and Cord bumpers.

A showroom announcing the arrival of the new Cord 810, billing it as a car five years ahead of its time.

A view of the Gentlemen's Roadster with top down. This was an attractive car—termed "a kingsize MG" by Gordon Buehrig, who felt it would have been a good seller if produced. The car still exists today.

The Cord display at Kunkel Motors in Brooklyn during the New York Auto Show week in 1936. Note the early use of a neon sign.

An early photograph of the prototype Cord 810, which still had the headlights mounted in the top of the fenders and did not have the production hubcaps. The picture was used in the first catalog.

The Cord 810 Sedan as introduced at the Chicago Auto Show in fall 1935.

A 1936 Cord 810 Phaeton, photographed by the dealer in Chicago for advertising purposes.

This picture, from a brochure issued in 1937 for the Supercharged 812, shows the Lycoming V-8 with supercharged intake manifold.

Features of the Cord 810 were outlined in this cutaway drawing from the second catalog.

The Cord Front Drive Embodies More New Features that contribute to safety and comfort, than any other automobile

1 Gear-shift unit, operated by remote control. In addition to the conventional low and second gear, the Cord has two high gears. The "extra high" gear is for high-speed cruising, with lower engine speed.

2 Power plant and driving unit are in one complete unit, including virtually all the mechanical elements of the car. Driving power is transmitted through bevel gears in driving unit, direct to front wheels.

3 Engine is an 8-cylinder, 90 degree V-type Lycoming, 3½" x 3¾", and developing 125 Horse Power at 3500 R.P.M. Rated Horse Power for tax purposes is 39.2.

4 The Cord Steering Wheel is of Tenite, in same color as exterior of car. Inside the steering wheel is a rim control for the horn, which can be operated by the thumb, without removing hand from wheel. Immediately beneath the steering wheel is the gear selector lever.

5 The front compartment is entirely clear due to the absence of gear shift and parking brake lever. Recessed floor pan adds to roominess.

6 Deep seat cushions with fully sprung and padded seat back placed at proper posture angle provide "relaxed comfort."

7 Rear seat passengers in the Cord ride on the same level as those in front, and well forward of the rear wheels. The low floor permits a natural seating position.

8 By eliminating transmission, propeller shaft and rear axle mechanism from beneath the car, "wind drag" from these sources is removed. Also there is no noise or vibration of mechanism felt by rear seat passengers.

9 The Cord body is built as a complete unit with its own frame, instead of having a separate chassis under the entire car. This body unit is wedged and bolted to the stub frame of the power and driving unit.

10 Road clearance of 9 inches obtained without sacrifice of passenger leg and head room; yet overall height of car (Sedan) is only 60 inches.

INDIANA-35
438 524

438-524

This view of the first prototype 810 Cord Sedan shows the headlights mounted high in the fenders, plus some trim differences.

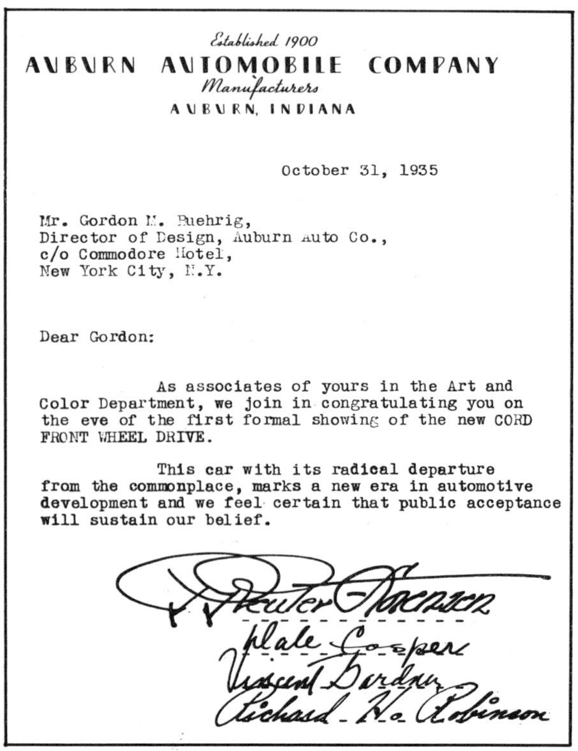

Established 1900

AUBURN AUTOMOBILE COMPANY

Manufacturers

AUBURN, INDIANA

October 31, 1935

Mr. Gordon M. Buehrig,
Director of Design, Auburn Auto Co.,
c/o Commodore Hotel,
New York City, N.Y.

Dear Gordon:

As associates of yours in the Art and Color Department, we join in congratulating you on the eve of the first formal showing of the new CORD FRONT WHEEL DRIVE.

This car with its radical departure from the commonplace, marks a new era in automotive development and we feel certain that public acceptance will sustain our belief.

The letter received by Gordon Buehrig from his compatriots at Cord, at the introduction of the 810 at the New York Auto Show in 1936.

The war years and industrial espionage | 10

Following conclusion of work on the 810/812 Cord project, there remained little to do at the Auburn Automobile Company. It was rapidly becoming apparent Auburn's ills constituted a terminal case and I started looking for a healthier company to work for.

The Budd Company

Finally, I made a good connection with the Budd Company in their Detroit office.

The Budd Company headquarters was in Philadelphia, Pennsylvania, and Edward G. Budd, Jr. was president of the company. There they built stainless steel trains and had other steel fabrication activities. The bulk of their automotive business was in Detroit and they had a factory there. This plant was divided into two activities; one building wheels and brake drums and the other producing automobile bodies. Some Studebaker body components and a good deal of the body tooling were still done in Philadelphia.

Their competitors in body work were the Briggs Manufacturing Company and the Murray Body Company, both located in Detroit.

My initial contact was with Ed Deisley, chief engineer for Budd, and he outlined Mr. Budd's plan. The company was only doing a fraction of the body work Briggs and Murray were doing. Budd had previously engineered and built the bodies for the "Airflow" Chrysler, DeSoto and Plymouth cars. The engineering was brilliant and the bodies were well built, but the public's refusal to buy the cars hurt the Chrysler Corporation and, in the trade, it also hurt the reputation of the Budd Co pany as well.

In 1936 the company was building the doors for Packard and the commercial bodies for Ford and Dodge. It bothered Mr. Budd that they were not building passenger car bodies. The LeBaron division of the Briggs Company had recently established an immpressive design department and this styling service was available to Briggs' customers. Much of Ford's styling at that time was done by this studio. The Lincoln Zephyr is one example.

Budd's plan was to establish an automobile design or styling studio in the Detroit plant with the anticipation such an activity would assist the sales department in obtaining passenger car body business.

After my interview with Mr. Deisley, he sent me to Philadelphia for an interview with Mr. Budd. This worked out well and I was employed to be chief designer and to build a small group for my staff. Mr. Budd was a delightful person and wanted our studio to be first class.

At first we had only a partitioned off area in the engineering department and we worked there for several weeks. Then Mr. Budd came out for a visit and he was so pleased with our progress he asked me to design a studio to be built in an area adjacent to the executive garage.

He requested I fly to Philadelphia in about four days with my plans. Our group enthusiastically accepted the assignment and instead of going to Philadelphia with just some sketches I had a small architectural model built of the design. Mr. Budd gave immediate approval and soon the work was underway.

The studio, when completed, was beautiful and Mr. Budd was generous in his praise. Mr. Deisley was also happy to have our department and we had fine cooperation with his engineering department.

One entered the studio through a curved hallway into a brilliantly lighted room with a full-sized car turntable at the front. Along the left hand wall was a display area with three turntables for one-quarter scale design models. On a raised platform along the right hand side were drafting tables for four designers with my desk at the front. This was a custom built arrangement with a control panel for spot lights, for activating the different turntables and with my own drawing board. All of what I have described occupied about half our floor space. The balance contained our woodshop, plaster casting facilities and a paint booth.

This view is looking into the entrance corridor of the Budd design studio in Detroit, which Gordon Buehrig and his staff created plans and a model of in just four days.

This view of the Budd design studio in Detroit looks from one of the designer's drawing boards across into the model shop, which was equipped with all the latest equipment for a design studio of the period.

One of Gordon Buehrig's design efforts at Budd was creating the design studio itself. This view is looking across the back of Gordon Buehrig's desk and drawing board, from which he could control an entire design presentation through use of the master switch panel seen at the right of the desk.

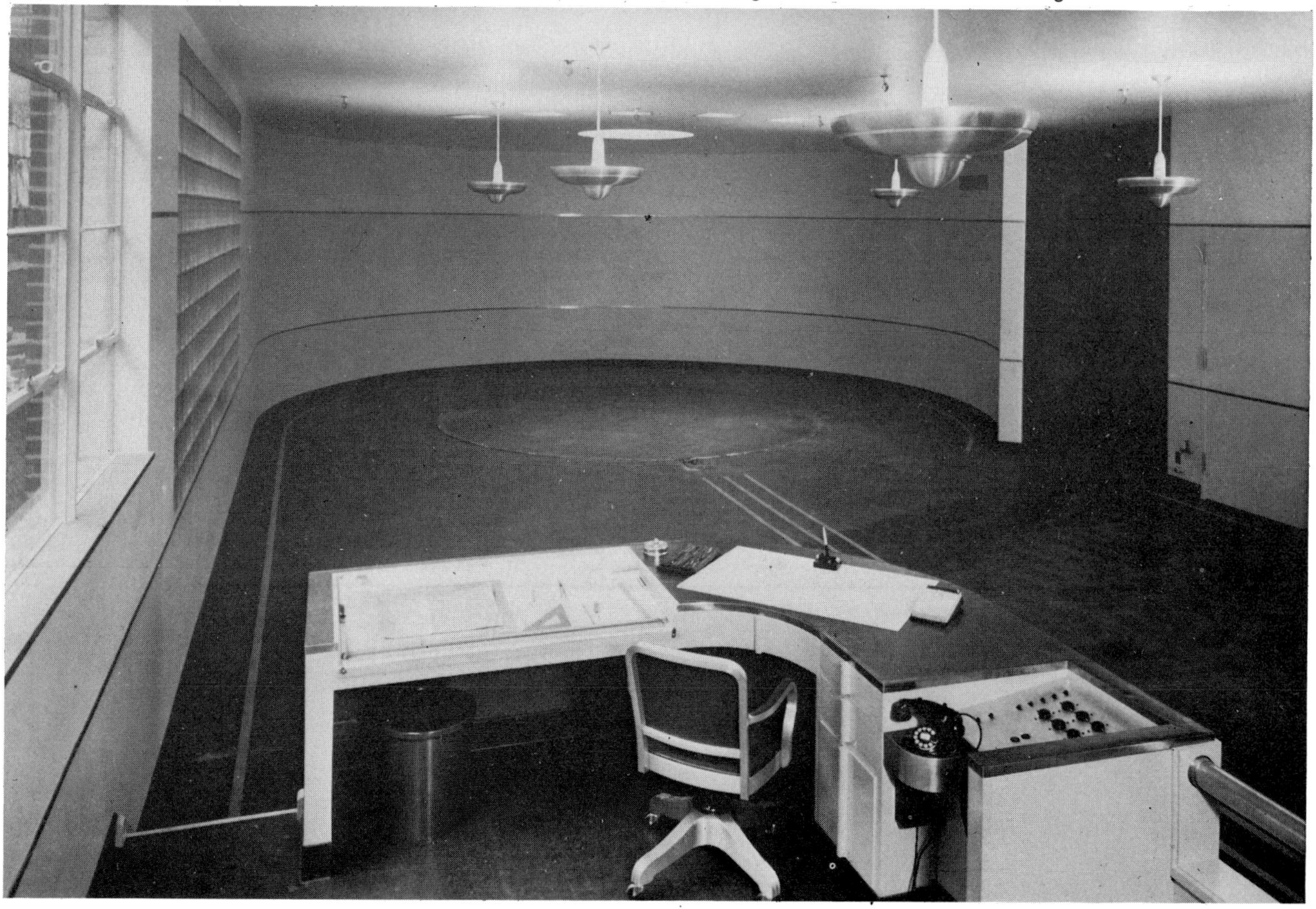

Although it was not apparent at first, I soon found out there was one powerful man in the company who did not like us. He was the vice president in charge of sales and it was for his department we were supposed to provide design assistance.

My first discovery of his attitude was at the time of the completion of the studio. Mr. Budd sent a PR man out from Philadelphia to plan a press party and reception for our customers to show them our studio. The plans were made but the vice president in charge of sales vetoed the party and ordered no publicity be given to the new design activity.

I worked on a number of designs as start-up programs for Budd. One idea was for an inexpensive car I called the "Wowser" which was truly an economy vehicle, right down to the interior; but the sales depart-

One of Gordon Buehrig's early design exercises at Budd was this two-door sedan which has obvious parentage in his 810 and 812 Cord designs. Note he is still interested in the hermatically sealed engine compartment and here has gone to a rear mounted radiator to achieve this.

The overall roof line and similarities to the Cord 810 are seen in this high 3/4 view of the scale model. This was one of the designs Gordon Buehrig had hoped Budd's sales department would sell.

ment never showed it to any potential customers as they were afraid it might offend the clients' engineering departments. If it had been shown we might have the Wowser today instead of the Volkswagen.

Month after month went by and we worked out many new designs, but no one ever saw them. This finally got to me and I wrote a letter to Mr. Budd complaining about the situation, saying I was wasting my time and his money. Mr. Budd's reply was short and to the point. He said that it was with deep regret he had to ask for my resignation. Thus, the termination was entirely my fault.

One of Gordon Buehrig's ideas while at Budd was for an inexpensive car that could be built with two basic body stampings, each making up half the body, as seen here in this Budd rendering.

The small economy sedan designed by Gordon Buehrig at Budd and called the "Wowser". Its body was to be stamped in two halves, with a seam down the middle of the roof.

Rear view of Buehrig's "Wowser" design at Budd. This was another of the designs he became disenchanted over when sales made no effort to promote it.

An interior mockup for the "Wowser" car at Budd. This was to be an economy vehicle, thus the sparse interior. Note early use of dash padding.

A Buehrig-designed cab-over-engine stake bed truck, done during his tenure as head of Budd design in Detroit.

I moved back to Auburn and tried to make a living as an independent industrial designer. I did some work for White Truck and quite a bit of work for King Seeley; I tried hard, but failed to make a go of it.

In 1940 I did a number of instrument panel designs for the King Seeley Company of Ann Arbor, Michigan, and we moved there. I was working for King Seeley when World War II started. This ended all automobile design efforts and I immediately got a job with Consolidated

After leaving Budd, Gordon Buehrig moved back to Auburn where he and John Rinehart made an attempt at consulting engineering. This instrument panel by Gordon Buehrig was an effort for the King Seeley Corporation, one of their customers.

Aircraft in San Diego, California, as a draftsman. I really enjoyed working in aircraft engineering.

I was surprised, both at Consolidated and at Goodyear, how well my experience in automobile engineering enabled me to do productive work in aircraft. While in San Diego I took a night school course in trigonometry because I never had it in school. In Akron I was actually working for a Detroit-based engineering company doing sub-contract work at Goodyear in tooling the F4U1 Corsair gullwing fighter for production.

The information we received from Vought-Sikorsky, which had designed the airplane, was in the form of orthographic drawings showing the parts projected in their "flying position" on the aircraft. These components had to be redrawn in more normal views for manufacture and had to be rotated in space to many different planes for machining operations. It was rather complicated and beyond the skills of some of our tool designers.

The geometric rotation of components in space is a common problem in automobile body engineering and my experience in this area was soon put in practice. With my then-recent schooling in trigonometry, combined with my body drafting experience, I was able to solve the problems for the tool designers. I became the mathematician for the group! It is strange how a person can adapt to a situation.

Thus employed, I spent most of the remainder of World War II.

Studebaker/Loewy

Near the end of World War II I returned to automotive design when I was hired by Raymond Loewy to manage his South Bend, Indiana, studio, which was then already at work designing the postwar Studebaker line. Most people think of the Loewy influence as something which came along after the war, but he was already well established as Studebaker's chief design consultant as early as 1939. He and his team were credited with the design of the 1939 Studebaker Champion, which had been well received.

Raymond Loewy, who hired Gordon Buehrig to head his South Bend, Indiana, Studebaker studio when tensions began to develop between Loewy and Virgil Exner. This picture was taken at that time. The design model in the picture shows the front of a sports car concept by Gordon Buehrig. Mr. Loewy liked it but Studebaker management was not impressed.

The Loewy design team at Studebaker is seen here at work during the postwar Studebaker design period. Virgil Exner is at right, leaning on the design board, and Gordon Buehrig is standing at center pointing to the plan mounted on the board by the window.

Virgil Exner, photographed by Studebaker photographer Bill Bradberry at the time the Studebaker controversy was developing.

My friend Virgil M. Exner had preceded me as head of Loewy's Studebaker efforts and I took on the assignment knowing full well I was walking into the middle of not only a design fight, but a management war as well. However, I was delighted to get the job and be back in the automobile business.

Exner had begun work on the postwar Studebaker as early as 1942-43 during war design slack periods. Exner had designs for the 1947 Studebaker well underway in 1944 when friction between Loewy and Exner brought me into the project as manager and Exner's boss. The confrontation between the two strong-willed men had been a long time coming and was understandable. Exner was an excellent designer, feeling a strong honest design would stand on its own merits. Loewy, on the other hand, was a promoter as well as a designer, and often made his point as much on his sales ability as excellence of the work. He was also willing to accept credit for his team's efforts in his own name as well. The combination of these two traits led to the Exner/Loewy split at a later date, which saw the fired Exner immediately go on Studebaker's engineering payroll and myself elevated to sole manager's post.

While Exner was still with the Loewy firm, our relative positions caused little or no problem. It was embarrassing to me to be in a position above him because we had known each other for many years and were friends.

Part of the time when I worked for the Loewy firm I was manager of the South Bend studio and Exner was my assistant. At other times Exner was manager and I was his assistant, as Loewy changed our titles. This made no difference in the way we worked. We had a mutual respect for each other and managed the office as a team.

One of my own personal designs for Studebaker was a sports car with a disappearing top.

Roy Cole was vice president of engineering at Studebaker and he admired Virgil Exner as a designer and also had ambitions for creating a Studebaker design department with Exner in charge. This plan included eliminating the Loewy organization from their Studebaker relationship. A political scheme developed in which Cole out-maneuvered Loewy and this put me in a bad situation. In spite of the fact that I did not know what

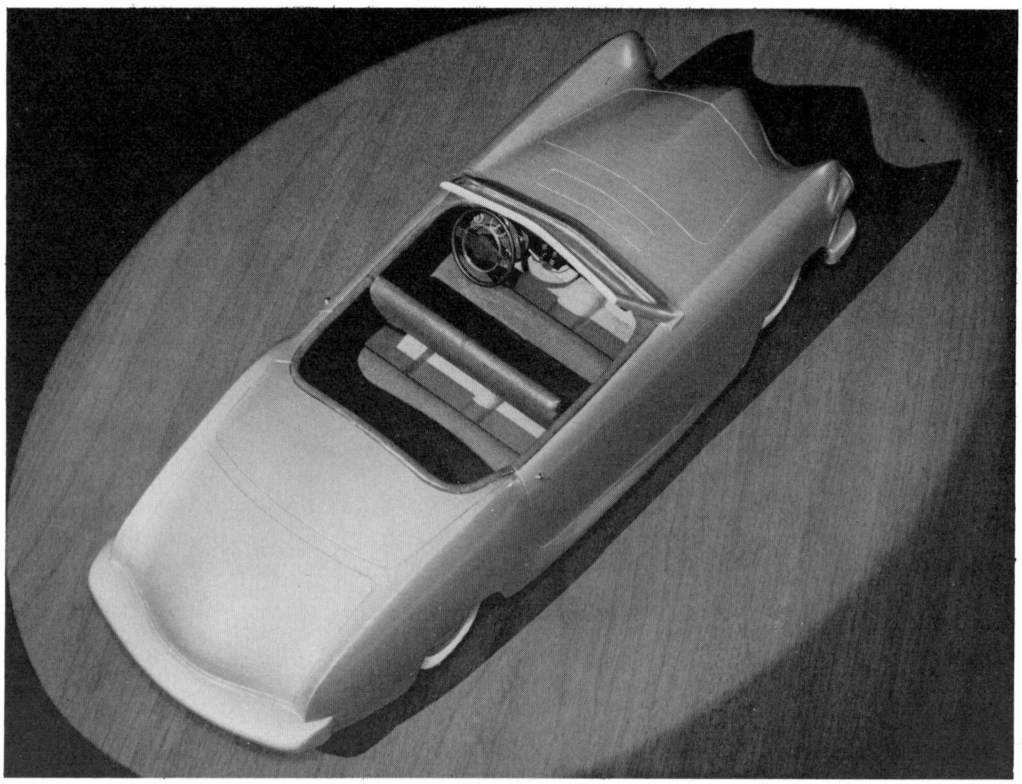

A Buehrig designed sports car with a disappearing top. This is another view of the car design model in the picture with Mr. Loewy on page 116.

was going on behind the scenes until the very end and was completely innocent, I don't think Loewy believed it.

Here is what happened. For several weeks it had been Exner's habit to arrive in his office about 11 a.m. He would mention he had been with Roy Cole, whose office was about two blocks away. This was a logical reason for his absence and I believed it. I never suspected Exner and Frank Alhroth were designing a car at night in Exner's basement. This was a scheme of Cole's—with the approval of Studebaker board chairman Harold Vance—to out-maneuver the Loewy team.

Some time later, Roy Cole called me to his office and Exner was there. Cole explained to me what Exner had been doing and that it was a special small car with a different package layout than the one he had

Designers at work in the Loewy Studio in South Bend. Reading from the left are Bob Kots, Vince Gardner, Bob Bourke, Frank Alroth, Jack Aldrich, John Reinhart, Virgil Exner, Raymond Loewy and Gordon Buehrig. John Bird (behind Loewy) and Larry Brom at the rear of the room, the two ladies, Virginia Spence and Nancy Spence were clay modelers and wood craftsmen.

given us to use. He said the one-quarter scale model Exner had designed had been sent to the Budd Manufacturing Company in Philadelphia to be produced as a wood model in full size. He also explained he was going to get rid of Loewy. He asked me to accompany Exner to Philadelphia to inspect the finished design model and to turn in my expense account to him. This was the first time I was aware of Roy Cole's dirty trick.

A few days later Studebaker's Board of Directors was called in for a showing of the Loewy design team's model, in full size clay, and the Exner/Budd full-sized wood model. Needless to relate, the Exner/Budd version of the 1947 Studebaker was the one chosen and all that remained was to finish details, design the interiors and solve some minor problems. It could have gone no other way, as the die models were already half finished at Budd for the Exner version.

Raymond Loewy was furious and he had a right to be. Paul Hoffman was president of Studebaker at the time and he, as well as Board Chairman Vance, must have been aware of the entire scheme. However, because of the great reputation of Raymond Loewy, the introduction advertisements gave Loewy full credit for the design.

A few days after the showing of the two models to management, Loewy fired Exner and Roy Cole immediately hired him and set up a design studio, with Exner in charge, out at the Studebaker proving grounds.

It was only a matter of a few weeks before A. Baker Barnhart, one of Loewy's associates, came out to South Bend and advised me to resign.

Very shortly thereafter I left the Loewy organization to begin work on the Tasco project.

Postwar Studebaker styling models on display for the Studebaker Board of Directors. Our first assignment on the postwar Studebaker program was for each designer to make an original design model. This photo shows the finished models which were being designed in the picture on page 118.

11 | Tasco

Someone once said, "Show me a man who never made a mistake and I'll show you a man who never did anything." When I think of the Tasco I repeat this to myself. It helps a little.

The Tasco, you might say, was my personal Edsel. It was not a very significant automobile in the broad scope of history, and since only one prototype was built it would seem that after a quarter century the car would be forgotten. But it still exists to haunt me.

The story began one day in 1948, while I was heading up the Loewy design team at Studebaker. I received a letter from Russ Sceli of Hartford, Connecticut. Russ said he and a group of his friends were interested in building an American sports car and would I handle the design work? I managed to get up to Hartford very quickly.

Some of Russ' colleagues were among the founders of the Sports Car Club of America and were talking, dreamily I thought, about an annual European-type grand prix race at a place I had never heard of called Watkins Glen, New York. Among the problems they foresaw was the lack of a suitable domestic vehicle to race in this and lesser events. Thus the idea for the Tasco, the acronym standing for "The American Sports Car Company." The whole idea had my immediate interest and I joined with the group, which was soon incorporated.

Altogether there were ten of us, and each invested $5000 in the firm. One member, Tom McKean, added a second $5000 later to help finish the prototype. No investment was ever recovered.

With the war just recently ended, most American companies were building basically prewar cars. They had no ready products that were truly modern or suitable for the purposes the infant SCCA had in mind. The consensus of the group was the proposed vehicle should be an open sports car with a special chassis and a fully American engine and drive-train. The prewar Jaguar SS-100 was admired by some and the early Morgan four-wheelers by others. A few said what they really wanted was a "king-size M.G." They were probably right and I was probably wrong, because I kept insisting on a closed car with a new type of top which I had in mind, employing twin removable panels on each side.

I argued the problem had a parallel in the private aircraft field. In the early days of flying, all private planes were open. The pilots wore leather jackets, helmets, scarves and goggles. They enjoyed the noise of the engine and the general excitement of that kind of travel. But in 1927 Lindbergh flew the Atlantic in an enclosed cabin, dressed in a business suit. From then on private aircraft gradually changed to the cabin type. I predicted the same thing would happen to sports cars.

Ultimately, as we can see now, I was right. But had I gone along with my associates' desires at the time, we might have been successful with an open car. After getting established we could have developed the more complicated closed variety.

I was able to convince them, or at least they agreed, that we should develop a cabin type sports car. I had only a few meetings with the group but many with Russ Sceli, who was president and principal organizer of the company.

Russ showed me a lot of pictures he had collected, including some design sketches by Claire Hodgman published in the English magazine **Motor.** One of these was a sports car with front fenders that turned with the wheels. Russ was intrigued with this feature and suggested it be an integral part of our design.

The more conventional fastback Tasco would have still been fresh styling in the mid-1960's. It was radical in 1948.

The two original design models of the Tasco. The more conventional fastback model was carried no further.

I was living in South Bend at the time and had a small shop in my basement where I could build scale models. It was there I made plans to develop the Tasco. I made two 1-8th scale models. The first was fairly well detailed, showing the windshield and window layout, the turning front fenders and the first concept of the top I planned to use. The second model was just a shape which I never finished in detail. This one lacked the turning fenders and was the one I personally preferred.

As I went into the turning fender problem, I became skeptical of the merits of the idea. I took both models to Hartford to show Russ, but his reaction to the second model was that it resembled the Buick fastback and was not sufficiently different to command a market. It would have been quite attractive though, looking a bit like the 1971 Camaro, and in actual size it was about the dimensions of the Corvette. But, in 1/8th scale its size and proportions were difficult to evaluate and Russ felt we should go all out with the more detailed, radical design with the turnable front fenders.

At this point I made a crucial mistake. I should have refused to retain the turning front fenders because I was aware of the problems they would entail. But at that time I thought I could work them out and, as Russ suggested, the car would surely be different. I returned to South Bend to design and build a quarter-scale model as per my original detailed design.

The finished quarter-scale model was completed in late 1948 and I took it to Hartford to present to the entire group. Not everybody was happy with it, but we were running short of time and the decision was made to go ahead with a running prototype.

Top view of the finished Tasco model. Note the instrument panel.

Gordon Buehrig is seen here with the finished model of the Tasco.

The engine in the Tasco was a modified Mercury, fitted with an Edelbrock overhead valve conversion.

We decided to use a Mercury engine with speed equipment, a modified frame, Mercury axles and drive line. Dale Cosper designed and engineered the frame to provide a compact layout. The chassis was shortened, the engine moved aft several inches, the "X" member removed. Frame stiffness was provided via tubular cross members.

Here our timing was unfortunate in several respects. The 1948 Mercury still had an L-head V-8 engine, though we boosted performance with an Edelbrock dual intake manifold and two carburetors and dressed it up with cast aluminium Edelbrock head covers, chrome plated parts and some striking paint. The transverse springs were really the same basic design as Henry Ford had used on the Model T, and a more modern chassis would have helped. Had we waited a year, for example, we could have used the 1949 Ford chassis with its independent front suspension and no cross member to remove to achieve the desired lowness.

As I had feared, the turning fenders produced many problems. Not helping the situation was my decision to have them molded of fiberglass, a radically new material in those days. Inboard, the fenders were attached to the brake backing plate. On the outside they were fixed to the axles. At the front there was no problem as the axle didn't revolve, but the rear axle is live and required a ball bearing support attachment. Fender width in plan view became a problem, even though they were as narrow as they could be and still clear the tires. They encroached on the hood and body, and everything became a little too tight. The better proportions I had shown on the first 1/8th scale model were not possible on the larger, more accurate layout.

We built the clay model for the fenders in my basement. The actual fiberglass versions were made up by my friend Howard Anthony of Benton Harbor, Michigan. Howard owned the Heath Aircraft Company, which years ago built the high wing, single-place Heath Parasol airplane, powered by a Henderson four-cylinder motorcycle engine. The Parasol sold for $1000, or $500 in kit form. After the war Howard began designing radios, hi-fi equipment and electrical testing units, and his resulting Heathkits are well known today.

The day Cosper and I visited the Heath plant to discuss the Tasco fender project, we found Howard working on molded fiberglass floats for converting light aircraft for use on water. To demonstrate the toughness of the new material he handed Dale a heavy mallet and invited him to hit a float as hard as he could. Dale did, and the mallet went smack through—a little embarrassing on both sides. Howard assured us the stuff would hold up better on the Tasco, so we told him to go ahead.

I knew the turning fenders, suspended as they were with the wheels, would add undesirable unsprung weight with a negative effect on the ride and handling. To offset some of this I decided to build the wheels of magnesium. Made by the Dow Chemical Company, they were well machined, beautiful and strong, but very expensive. However, their light weight fully negated the added weight of the fenders.

I continued to make compromises for one reason or another in developing the final design. Unfortunately, finances required the elimination of the disappearing headlights to simplify construction and cut costs. Originally I had planned to have a bumper nose and tail section of pneumatic construction, with the units built like a tire with an inner tube. A wide steel plate would act as the "rim." It would have been novel and functional, but again, the cost of the molds made the idea prohibitive.

Later I planned to expose a segment of the spare tire to act as a bumper. The spare was mounted in a horizontal position and was strong enough to function in that way. I finally decided to enclose it behind a hinged, plated bumper.

One of my more serious mistakes, which largely contributed to the broken-up lines of the finished car, was the conflict of the daylight openings or glass areas with the overall design. At the rear, for instance, where a transparent acrylic bubble housed the licence plate, tail and turning lights, the lines didn't "read" properly. In the quarter-scale clay, these problems were not as apparent, though they forcefully appeared in the prototype. By then we were short on both time and money.

In the end all these difficulties were rather discreetly ignored. We bought a new Mercury in South Bend, discarded the body, modified the frame as per the plans previously laid down, and provided the chassis with lights, a seat and other necessities to make it driveable.

Together with Vic Simney, a young model builder who had helped me construct the vehicle, I drove the bare chassis from South Bend to the Derham Body Company in Rosemont, Pennsylvania, where Enos Derham had agreed to build the body. Driving the chassis was a great experience. Dale Cosper's taut frame returned excellent handling characteristics. Recalling this emphasizes my belief that had we fitted it with a simple open car body we would have had a fine sports car on our hands instead of a disaster.

The cost of building the body went far beyond our anticipated figure, even though Derham did the work at little or no profit. When the car was finished, it was obvious that extensive development work was

Gordon Buehrig in the roadable Tasco chassis, dressed for the drive to Philadelphia.

The Tasco chassis as driven to Philadelphia and the Derham shop.

The Tasco under construction in the Derham Body Company shops, Rosemont, Pennsylvania.

still required, and I became convinced the problems inherent in the turning fenders could never be completely resolved. Steering at high speed was sensitive to cross winds because of the fenders, but precise steering and high speeds are prerequisites in any sports car.

But all wasn't discouraging. I was sure my top design with the removable panels was an important invention, and the body interior had a number of unique ideas of merit. The general feeling one had inside the car was good, much like sitting in a two-place airplane.

The instrument panel was located high on the cowl and the emergency brake was only a short distance from the steering wheel. There were two gasoline tanks and an aircraft-type switch to go from one to the other. The driveshaft was covered with polished stainless steel, topped inside the cockpit by a leather-covered box enclosing the wiring and fuel lines. This was probably the first instance of using the driveshaft tunnel as a design feature.

By the time the prototype was on the road, our hastily contrived Tasco Corporation was out of funds. I spent about six months with the car trying to interest some company in producing it but always with the understanding we would build a second prototype using only the good features from the first one.

I showed the car to Ercoupe in Baltimore, Beech in Wichita, Consolidated in San Diego, Lockheed in Burbank and many other aircraft companies. Because of the aircraft-type design of the Tasco and the

Steering column and foot pedals of the finished Tasco.

Completed instrument panel, with an intentional "aircraft" look.

large production capacities these companies now had sitting idle at the war's end, I reasoned they would be good prospects. However, the answer was always discouraging.

When we realized we were licked we dissolved the Tasco company. I was in the red for $2000 in travel expenses, in addition to my investment, and since the car had little resale value it went back, I believe, to Derham, who sold it. I made an agreement with the company to resolve my expense billing in return for a clear claim to the top idea. Later I went ahead at my own expense and obtained a patent on it.

In 1949 I joined Ford, taking charge of one of their styling studios. The top patent was granted to me on June 5, 1951. I discussed it with E.C. McRae, then head of the patent department at Ford. He said the patent was entirely my property since it had been invented and the patent applied for before I started working for Ford. Furthermore, I was free to try to sell it to the company.

Late in 1951, when I was working in the Lincoln Division as a member of the Product Planning Department, I made an artwork presentation of the idea to Richard Kraffe, president of the Lincoln Division. He saw no value in it, so I gave up my efforts to sell it for awhile. Dick Kraffe later became president of Ford's Edsel Division.

The finished Tasco, carrying a 1947 Indiana license plate.

In 1955, Ford introduced their two passenger Thunderbird sports car. I bought one and engineered a Buehrig top for it. Using one of the removable fiberglass hardtops, we reworked it to display my patent. Stanley Brom and I made the tooling for the parts in my garage and Vince Gardner built the top.

By this time I was Station Wagon Planning Manager in the Ford Division. The head of Product Planning was Chase Moresy, under Division president Robert S. MacNamara.

One day as MacNamara was leaving for lunch I asked him to look at the top, and he seemed impressed. Through Tom Case, Thunderbird planning manager, Ford patent counsel Jack Faulkner and someone in the legal department, I had a contract drawn up for a sale of the patent to Ford. I signed the agreement and it was put on the agenda for a meeting of the Ford company board of directors.

I was beginning to feel pretty optimistic. It had been my hope, ever since the Tasco folded, to make enough money on the top to repay each of the original investors the money they had lost.

The day of the board meeting passed and I heard nothing. The next day I inquired what happened and was told MacNamara had removed the item from the agenda before the meeting. I was also told that since he had decided against the top, I should just forget it!

Sometime later I inquired through proper channels if I might try to sell the idea to another automobile company without jeopardizing my position or my annual bonus. Ford said I was free to do this. I called an old friend, Bart Cotter, who at the time was director of engineering at Fisher Body and had been assistant chief body draftsman at Auburn in 1934 when we did the Cord 810. He set up an appointment for me to demonstrate the top to engineers from Fisher Body and also a number of designers from GM Styling at the GM Tech Center. The result, as Bart reported several days later, was GM was not interested.

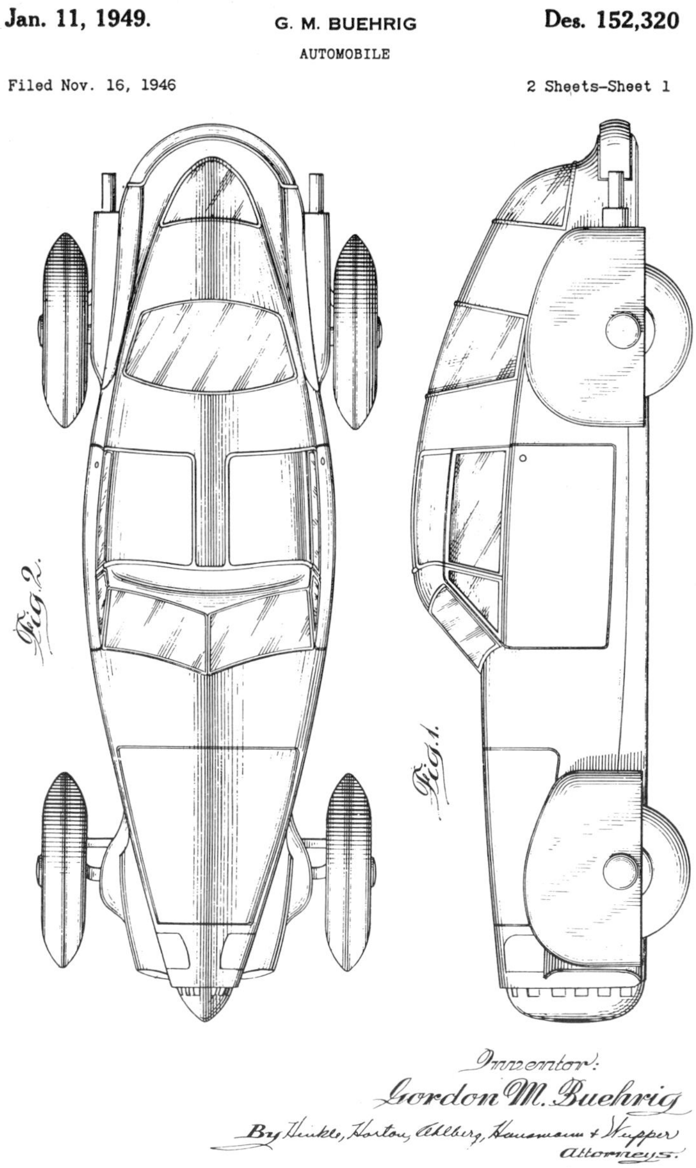

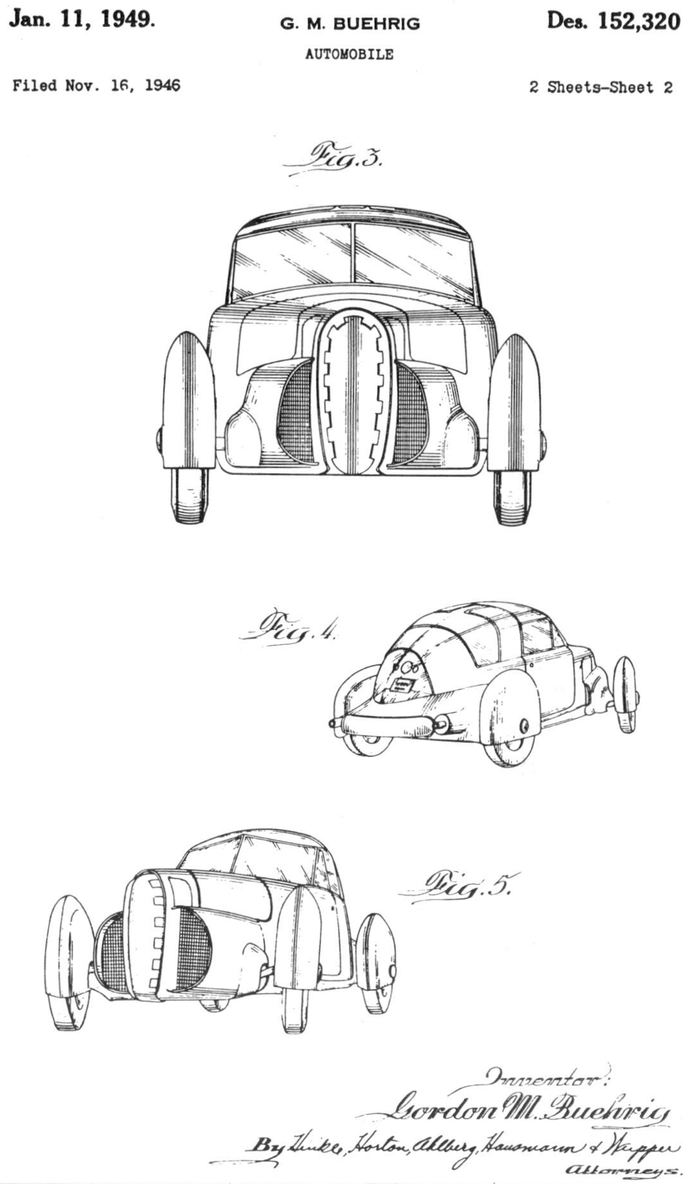

Patent drawings for the Tasco.

Gordon Buehrig here shows his daughter, Barbara, how to install the clear plastic panels in his Thunderbird's removable hardtop.

With panels removed, the top has all appearances of a modern Corvette.

Patent drawings for Buehrig Vehicle Top.

June 5, 1951 G. M. BUEHRIG **2,556,062**
VEHICLE TOP CONSTRUCTION
Filed July 13, 1948 3 Sheets-Sheet 1

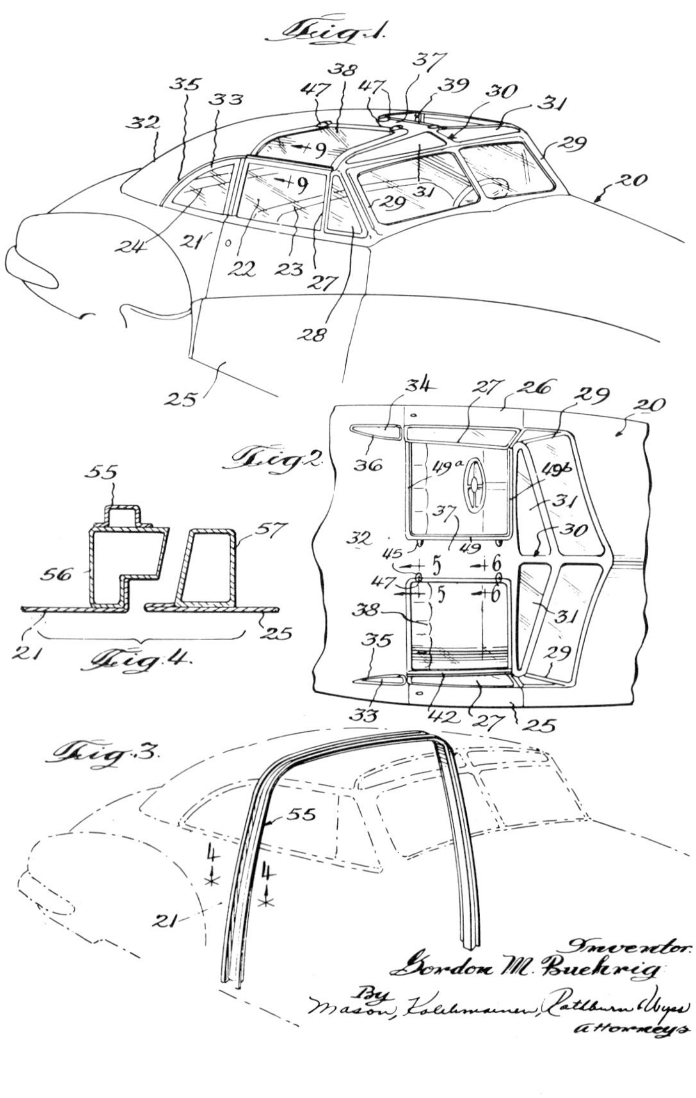

June 5, 1951 G. M. BUEHRIG **2,556,062**
VEHICLE TOP CONSTRUCTION
Filed July 13, 1948 3 Sheets-Sheet 3

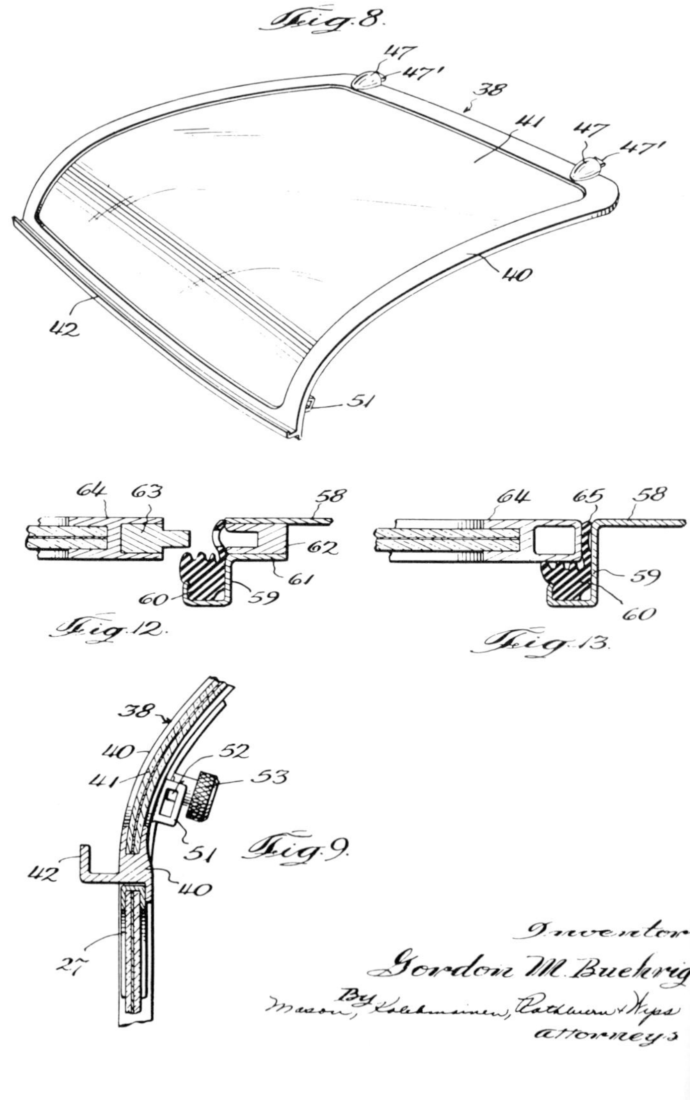

My next attempt was to sell the top to Chrysler. Another friend, Virgil Exner, who was then a vice president of the company and director of styling, arranged a meeting for me at their styling center, but the result was the same as at GM. Through another friend, Bill Gobell, I demonstrated the top at American Motors with the same negative result.

I even tried it out on the Golde Company of Germany, which builds sliding roof units for Mercedes and others. The company had a small shop in Detroit where they did custom top conversions. I showed Mr. Golde the top when he was in town on a business visit, hoping he would build one on a Mercedes for the European auto shows. I calculated if American auto executives saw it in Europe, they might be impressed. This meeting was also unproductive.

I had just about forgotten the whole idea when, in October 1967 I received my copy of **Road and Track** in the mail with a picture of the new Corvette on the cover. There, before my incredulous eyes was the old Tasco/Thunderbird top I'd been trying to market for nearly 20 years! The description of the car which featured the top was, in my opinion, an infringement of my patent.

My first thought was the 17 year life of my patent had expired and GM had waited until its death to use my idea. However, I found a copy of my patent in my files and it was not due to expire until June, 1968.

I wrote to General Motors and finally hired a lawyer to help me make a settlement with them. I was not experienced in this type of negotiation and probably did not handle the alledged infringement as well as I should have. I asked for $100,000 and had I received it, the original stockholders of Tasco would have had their investment returned.

In Summer 1968, GM sent Herbert Furman from their Patent Section to Los Angeles to make a settlement if possible. He was quite friendly and explained their position in detail. He told me the amount he was authorized to offer me for the dead patent and said if I wanted more I would have to sue GM. He did not pressure me to take the settlement and said if I decided to sue and won the law suit, we would still be friends. He said it was purely a business gamble I would have to take, matching my resources against theirs in a litigation which could drag out for several years. Since I had been unable to retain a lawyer to handle the case on a contingency basis, I decided to accept his offer.

Recently I bought a Corvette, and whenever I drive it I think of the Tasco and the sports car that might have been. The Corvette top is, of course, much better engineered than that of the Tasco or my 1955 Thunderbird. However, it had four or more years of research and development at GM and is really excellent. It is probably the convertible of the future because air conditioning makes the rag top literally passe.

There were three features, though, on my Thunderbird top which I think had an edge on the Corvette. First, I had two interchangeable sets of panels, one of painted fiberglass to match the roof color and the other formed of clear acrylic. During the two years I drove the Thunderbird I seldom used the painted panels because the transparent ones were so much fun. Second, the Corvette has a makeshift arrangement for panel storage, whereas my T-bird had a sort of "file case" located above the differential in the trunk area which nicely accommodated all four pieces. Third, my roof had a wider fixed center section which added strength and safety and rendered the removable panels smaller, lighter and easier to handle and store. It also gave the interior more of an aircraft feel and appearance. My original patent also included a roll bar structural member built in, which I believe was the first such description of this safety feature.

One other aspect of the Tasco program was eventually of benefit to Ford. The U.S. Rubber Company of Mishawaka, Indiana, had a then-new plastic material called ABS (for Acrylo-nitrile-butadien-styrene) which was vacuum formable when heated. I made the quarter-scale design model of the Tasco from this material, having the parts formed by U.S. Rubber.

In 1949, when I was in Styling at Ford, one of my weekly duties was to attend the Body Engineering meetings where the construction of a new series of bodies was being developed. The problems of construction were described in these meetings by using perspective drawings and regular orthographic layouts. My purpose in attending was to catch any construction details that might alter the already-accepted body contours or that might present unacceptable joint lines. Since this rarely happened I found the meetings exceedingly boring and I would concentrate on every word, trying to keep from falling asleep. The meetings were scheduled for every Monday at 1 p.m. and I would drink several cups of black coffee and take a No-Nod pill before each session.

Sheet metal fabrication of an automobile body is quite complicated and sometimes I found it difficult to have a clear picture of the construction details under discussion. One day, after a meeting, I suggested to Henry Grebe, one of the body engineers, that they use three dimensional models to illustrate their construction proposals. He informed me it couldn't be done, as they had tried it with a model of a Lincoln windshield pillar. It had taken three months to build the model and I noted it was made of steel parts which had been formed over Kirksite dies. I suggested vacuum forming the parts with ABS, as I had done with the Tasco model.

Grebe was skeptical, having never heard of the material, but he did let me borrow the wooden patterns to demonstrate the process. I shipped the forms to U.S. Rubber in Chicago. George Martin, supervisor of Ford's Styling shop, and I flew over to see the parts made. They had the patterns ready and all 13 pieces were made in half an hour. I brought them back from Chicago, trimmed and assembled them at home and took them in to show to Henry Grebe. At first he wasn't impressed, but later the idea was put into operation by Manufacturing Engineering in a special shop at the Rouge plant.

In 1952, when I was assigned to Special Product Operations as body engineer on the Lincoln Continental Mark II program, I had a fine opportunity to fully develop the use of vacuum formed parts in a complete body program. Today there is a large shop at Body Engineering to do this work. It is used constantly to study construction of all new bodies. This engineering tool which is so successful today really dates back to the quarter-scale Tasco design model which, to my knowledge, was the first automotive application of ABS vacuum molding.

In retrospect the Tasco holds mixed memories for me. In itself, it was a failure, but many better ideas have also failed. While I have always regretted the losses it sustained, the fun of working on it and some of the concepts it fostered which later saw volume production make up for much of my regret over the car itself.

12 | Ford Motor Company

On August 15, 1949, I started work for the Ford Motor Company and retired from there 16 years later on August 15, 1965.

My assignment for the first two years with Ford was as manager of the Body Development Studio in Ford Styling. At that time there were six studios in Styling. They were Ford Interior Studio, Ford Exterior Studio, Lincoln-Mercury Interior Studio, Lincoln-Mercury Exterior Studio, Ford Truck and International Studio and mine, Body Development Studio.

The Body Development Studio had the responsibility of designing the convertibles, station wagons and any other special models for Ford, Lincoln and Mercury after the basic design for these cars had been completed on sedan full-sized clay models. In developing the other models we were supposed to incorporate in our designs a maximum amount of interchangeability with the approved sheet metal from the sedan models. The purpose was to hold down tooling costs, especially on the shorter run specialty models.

There were two accomplishments during my tenure in Ford Styling that deserve mention. The first is the story of how the 1951 Ford Victoria came into being.

In 1949, a model generally known as a "Hard Top Convertible" was a popular design in several General Motors line cars. The name was a misnomer in that it was not a convertible at all but the side windows and rear quarter windows were the same design as on the convertible. The

One of the first jobs Gordon Buehrig had after joining Ford in 1949 was a crash project to add a hardtop coupe to the 1951 Ford line. His restrictions included no change in tooling and the car had to be based on the current convertible model. The 1951 Ford Victoria was the result—a very successful automobile.

This artist's rendering shows the lines of the Buehrig-designed 1951 Ford Victoria hardtop coupe. The car came so late to the line, it was not included in the 1951 Ford announcement catalogs. This drawing is from the special Victoria brochure, Form No. 7289, issued in January 1951.

basic design probably owed more to the old "California Top" conversions available for touring cars in the 1920's and 1930's to make them more all-weather vehicles than it did to the rag top. The result was the same. With the windows lowered there was no window frame and no "B" pillar above the belt line.

Ford sales were suffering because they didn't have this model in their line and the Styling Studios had only designed one for the 1952 series, which was to be the next major body program.

John Oswald was the body engineer for the Ford line, also the superivsor of Styling and the man who hired me. One day in 1950 he came into my studio and asked if I could design a "Hard Top Model" from the current convertible with no changes in tooling, which would allow Ford to have a "Hard Top Model" in 1951.

We had available a quarter-scale drawing of the convertible and a quarter-scale model to work with. I drew the design lines on the quarter-scale print and in three or four days we had completed it as a quarter-scale design model.

It was accepted and went into production as a 1951 model called the Ford Victoria. It was so successful the already-approved 1952 design was scrapped and a new one copied from the 1951 design was released for production.

Shortly after this car was on the market I had a call from Jack Keegan, an old friend and co-worker at Dietrich in 1926. The reason for his call was interesting. He had just seen one of the new 1951 Ford Victoria's in a dealer's show room. He said he had heard I was back in Detroit and working for Ford, then asked, "Gordon, did you design the Ford Victoria because it looks to me like it has a Buehrig Roof line?"

In a small way, this proves some automobile designers, like some architects, have an individual feeling for lines and surfaces which trademark their work.

My second accomplishment at Ford Styling was the design work my staff and I did on the 1952 Ford and Mercury station wagons.

In 1951, the combined sales of Ford and Mercury station wagons was only 28,500 units. On the other hand, the bodies for the 1952 series were in production for three years and their sales for the first year were only limited by their capacity to produce the cars.

Earl McPherson gave me the station wagon assignment and also his personal ideas for the solution to the problem. He gave me a description of what he expected the new Ford station wagon to be and it was a perfect description of the wagon in current production by Chevrolet. It was to be an all-steel four-door wagon with door panel and quarter-panel stampings simulating the structure of a wooden body. All of these parts required completely unique tooling. The body was to be decorated with imitation wood decals over the steel stampings.

I was much in favor of going to the all-steel wagons but I didn't want to go the Chevrolet route. To do so would be a violation of my design ethics. I wanted the steel wagons to look like steel bodies. I also wanted to include a two-door wagon and a light truck version along with the four-door model. The sales department wanted one model with wood trim. We added that too, but with the wood only as decoration and not as imitation structure.

The sheet metal from the sedan which we inherited was not ideal for our purpose but I was determined to use the doors if possible because the tooling cost of doors is very expensive. I had orthographic drawings of the sedan in quarter-scale and in full size. The problem area was in the top line of the door where it proceeded aft from the windshield in a rising curve until it reached the rear of the front door. On the rear door this line curved downward from this point. It was easy to see the four-door wagon could use the front doors from the sedan but it would require new rear doors.

On the two-door model we had a tempting situation to cheat so we could use the sedan doors. The door on the two-door sedan was about four inches wider than the front door on the four-door sedan. The top line of the door on those extra four inches had started down hill, but only a little bit. It was just about a quarter inch. If we used the sedan door the result would be the parallel space between the top of the door and the bottom of the drip molding would not be parallel at that point. We decided to do it.

The car went into production and this detail probably saved the company $200,000 in tooling and a small saving to the customer because it provided a lower base for establishing the selling price. I doubt if any customers and probably few in management ever noticed this detail. The body shell was used on the 1952, 1953 and 1954 Ford and Mercurys. This interesting detail can be checked on any of the two-door (Ranchwagon) vehicles produced during those years.

On the 1951 Ford nine passenger station wagon, the rear bench seat was removable. The center seat folded into the floor so a flat floor aft of the front seat was available if you wished, but the rear seat had to be left in the garage. This was the standard arrangement on nine passenger wagons at that time.

On the line of wagons in our proposal, the two-door (Ranchwagon) was a six passenger vehicle, the four-door wagon was available as a six passenger and a nine passenger and the wood trimmed (Country Squire) version was only available as a nine passenger wagon.

Gordon Buehrig and his design team did the 1952 Ford all-metal station wagon line and literally turned Ford station wagon sales around. It was Ford's first all-metal wagon.

I wanted the nine passenger models to have an interior arrangement whereby all passengers would face forward, as they had in the 1951 Ford wagons, but with a seat design layout that would eliminate the problem of taking out the rear bench seat and leaving it in the garage when the vehicle was used with a flat floor aft of the front seat.

This required invention and I assigned the problem to George Golubics. It took him about a week to arrive at a solution satisfactory to him. He showed it to me, I approved it, and had it approved by body engineering. Ford took out a patent on George's invention.

With this invention Ford introduced the first nine passenger station wagon wherein the rear seats folded into a flat floor configuration. On the six passenger models the engineering of the seats was very simple.

The two-door model (Ranchwagon) went on the market as one of the lowest priced models in the Ford line, yet with a good profit margin. In 1951 the station wagon, also a two-door model, was the highest-priced car in the line. The Ranch Wagon quickly established itself as the best car in the Ford line for low depreciation. The 1951 station wagon had the highest depreciation of any Ford.

With the introduction of the 1952 models, Ford immediately dominated the station wagon market of the American automobile industry. By 1954, Ford station wagon sales reached 140,000 units per year, up from 28,000 units in 1951.

13 | Continental Mark II

Early in 1952, the Ford Motor Company top management expressed an interest in building a new Continental to be a prestige car. Their desire was to produce an automobile which would, without question, replace Cadillac in the minds of the American public as the finest American prestige car.

The formal proposal to the Executive Committee was written and delivered by Douglas Holmes. Doug later told me much of his proposal came from my earlier material. The general concept was that the Mark II should be the best car in the world. A profit to the company should not be a consideration, but it would be desirable to break even. The real benefit would come through the umbrella effect on the rest of the Ford automobile lines as a result of the Mark II's image as the world's best car.

The Continental Mark II brought two advances in car design that improved safety and one of them had a direct effect on appearance. This was the thin windshield post. This structure in auto body engineering is called the "A" pillar and its width in cross section is a governing factor in the driver's scope of vision.

In 1926, when I had worked for Ray Dietrich in his custom body shop in Detroit, he had invented a "Safety Windshield Pillar" made of cast bronze and narrower than the distance between a man's eyes. His patent claim was this eliminated the blind spot and anyone building a windshield pillar that narrow violated his patent.

In 1952 windshield pillars had become quite thick. This component at that time on our Lincoln car was just as big and had the same construction as the Fisher bodies, which reflected the fact our chief body engineer in the Ford Motor Company had recently come from Fisher. A large windshield post obstructs vision and consequently, from a safety standpoint, the narrower the pillar, the safer the vehicle.

The second safety feature was a bi-product of our package study. Reinhart's desire to have a low automobile required a layout with a low floor. Copp knew a quiet automobile with good handling characteristics required a strong rigid frame with a deep cross section. The answer was found in building the frame outboard of the floor and nesting the exhaust pipes along side the frame members. The resulting frame design could be called the prototype of all the frames used today, except it was much stronger. This added strength added weight to the frame and lowered the center of gravity. Bob Reganhart, Harley Copp and Bruce Clark were responsible for its design. The safety aspect came from the location of the heavy frame and its resistance to impact along the side of the car.

The contribution to automobile art of this frame design was the wide, low package and the new proportions it gave.

One part of the Continental program I especially enjoyed was building our "Cobbled Cars." We had them constructed by the Hess & Eisenhardt Company of Rossmoyne, Cincinnati, Ohio. In all, six cars were built there and one of them was a convertible. A separate group in the Continental program also built a convertible hardtop model, but the one we built was a conventional one, simply called X-1500 Mechanical Prototype Car #4. This is not to be confused with the single Continental Mark II convertible, which was conceived by John Reinhart and constructed under the direction of Enos Derham at the Derham Custom Body Company in Rosemont, Pennsylvania. That car was first shown at the Texas State Fair in Dallas in October 1956, and then presented to Mrs. William Caly Ford.

Gordon Buehrig is seen here with the first initial sheet metal mockup of the X-1500 "Cobbled Car", which was built on the new Continental Mark II chassis.

Front 3/4 view of the Continental Mark II "Cobbled Car" prototype. Note the chrome initials WCF at the top of the grill, for William Clay Ford.

The Continental Mark II prototype hardtop coupe version built by Hess & Eisenhardt is seen here beside a stock 1953 Lincoln hardtop coupe. The four inches taken out of the car by use of the new Continental Mark II chassis can easily be noted.

The completed Continental Mark II prototype test car in hardtop coupe version. It shows its family lines, having been made up from sectioned body panels from the 1953 Lincoln.

Our six "Cobbled Cars" used the massive "cow belly" chassis frame, floor pan and all other chassis components of the Continental Mark II. The bodies, however, were constructed from current Lincoln stampings which were cut up and pieced together in a plan which gave the car the proportions of the Mark II. The cars were lower, the hoods longer and the greenhouses shorter than the stock Lincoln. It was my only chance on the entire program to do a little styling. The exercise reminded me of my Model A Ford which I redesigned and built while working at Duesenberg in 1930.

The Cobbled cars were quite handsome, entirely because of their exciting proportions and their lowness. We built the cars around Harley Copp's steel rail chassis frame, which angled down quite low between the wheels. By placing the seat frames deep between the chassis frame rails, we were able to take three inches from the car's overall height and had no reduction of headroom inside the car.

The convertible cobbled car—we had planned a convertible in the Continental Mark II line—proved in the structural test laboratory to be the best convertible they had ever tested.

For our convertible program we purchased a number of different makes to study. The one that was outstanding was a Mercedes-Benz Model 300S. The car was absolutely quiet with the top up or down. We studied it carefully to learn the secret of its marked superiority and we found it. This car had been built with compliance (flexibility) in a number of areas where you would normally expect rigidity. A structure that is subject to torsional loads is far stronger if it has a carefully controlled compliance engineered into the design than if it is designed for complete rigidity. We applied the lesson learned from the Mercedes and the result was gratifying.

This 3/4 front view shows the "Cobbled Car" Continental Mark II prototype in the Hess & Eisenhardt factory upon completion. It was built using sheet metal from the standard 1953 Lincoln.

Rear view of the convertible "Cobbled Car" Lincoln prototype following its completion by Hess-Eisenhardt.

A significant bi-product of the Continental Mark II development program was the invention of what has come to be known as Three Dimension Engineering. On June 24, 1953, we displayed a full-sized clay model of the Mark II to the executive committee.

The work that had to be done from this point by Engineering, Purchasing, Manufacturing and Accounting was all carefully scheduled and integrated. As a result, departments other than Engineering required information from the engineering sections immediately. The only development work completed at that time was the construction of a mechanical prototype—the "Cobbled Car."

This car consisted of an experimental Mark II Continental chassis fitted with the body and sheet metal constructed from reworked Lincoln parts. Development of this car provided the Chassis Section with a fairly accurate and complete list of parts and prototype design information. However, the cobbled car was practically useless to the Body Section for prototype design information, except in the area of seats and underbody.

The Purchasing Department had to source all the body parts for the new car and, because of the planned limited production, this required mostly new vendors. To do this, they had to know the size, shape and form characteristics of each individual body stamping.

Rear view of the Continental Mark II quarter-size clay model. Note scribe line on the roof at back edge of door—probably a line drawn during study and work on the retractable hardtop version.

Front 3/4 view of the quarter-size clay model of the Continental Mark II. Note grill and side trim have not been finalized.

A low angle shot of the Continental Mark II Clay Model II, photographed at Ford Design Center.

The Manufacturing Department had to process the unique vehicle and, again, the limited volume produced a new set of problems. Manufacturing planning required a breakdown of parts included in each major assembly.

The Accounting Department had to develop a design cost analysis and a corresponding tooling cost budget for planning purposes. To do this they needed essentially the same information as the Purchasing Department.

The Body Engineering Section, which I headed, produced the required information by developing a completely engineered 3/8 scale model Continental Mark II in a period of three months. The results of this program were only possible because of the invention of Three Dimension Engineering.

Four copies of the model were constructed so Purchasing, Manufacturing, Accounting and Engineering could each have one. The parts for the model were vacuum formed of 1/8 inch thick sheets of ABS plastic, which has already been mentioned as having been used to make the design model of the Tasco.

Continental Mark II Clay Model II is seen here on the turntable outside the Ford Design Center. Note emblem on the spare which was not used.

Clay Model II of the Continental Mark II is seen here on the turntable at Ford Design Center. Both shape and trim are nearly finalized in this picture.

A side view comparison between a 1948 Lincoln Continental at front and Continental Mark II Clay Model I at rear.

To make the models, the Drafting Section first took templates from the full-sized clay model and made a full-sized body surface development. They had to do this anyway for their work. The various surface templates were then reduced photographically to 3/8 scale. From this information a 3/8 scale drawing was obtained and a 3/8 scale model built in clay.

A plaster "master model" was cast from the clay and the ten-inch section lines were scribed on the surface. The outline of each body surface panel, such as the doors, deck lid, hood, etc. were added.

Tooling for the body parts was developed by first casting a female mold from the master model. This mold served as the tool in which the plastic was vacuum formed. The plastic sheet was heated in an oven before being formed. This softens the material so the vacuum process will give it the correct shape.

After the outer panel was made the female mold was used as an armature for the inner panel design. This was made by clay modeling the configuration of the structural part (the inner panel) directly into the female mold. In some cases the engineer made a rough free hand sketch of the structure for the clay modeler to follow. Sometimes he made a rough orthographic drawing of the part. In any case, he designed, altered and refined the structure in the clay model stage. From this model a female plaster mold was cast and the plastic inner panel was vacuum formed to its configuration.

After the inner and outer panels were carefully trimmed, they were bonded with an adhesive and clamped into the female outer panel mold to hold them in the proper position during the curing cycle.

After each part or assembly was completed, it was photographed. The photographs were mounted on a piece of illustration board and in one corner there was a small scale orthographic drawing of the part showing the overall dimensions and the depth of draw of the stamping. The name and part number were also shown. From this, reproductions were made for the Picture Parts Book.

A typical part, such as a deck lid, is normally shown only on a full-sized drawing and it is probably fifty-six inches wide by eight to ten feet long. Such drawings are called "bed sheets" and they are a nuisance to refer to when the information you want is only the size, the shape and the draw requirements.

Our Picture Parts List consisted of a loose leaf binder with 8x10 inch pictures of every body part on the vehicle. It was a great time saver for everyone. An additional loose set of "stampings" were also supplied Purchasing for discussion of individual parts with vendors.

Charles deVoss, manager of the Production Department for the Continental Mark II said, when I delivered his copy of the model, "Gordon, for my part of this program, this model is worth $50,000 to me today!" It proved a great money and time saver to all departments.
a procedure common in the automobile industry today.

In the engineering of the Mark II Continental I remember the great pains we went to in assuring the car would be built as fine as possible.

Charles Russel, under Reinhart's supervision and suggestions, worked out a beautiful instrument panel. It was very mechanical looking but it had a "jewelry" look about it because of the fine execution of detail. The mock-up panel Stan Brom built was flawless. Our problem in engineering was it could not be built by standard automotive practices and be that good.

Another detail was the hood ornament. An ornament or emblem was something we had let ride without a final decision when one day upper echelon management said they wanted to see what we had in mind . . . tomorrow! Bob Thomas worked up a design sketch in a hurry for the board meeting and it was approved unanimously. Charlie Pfaneuf, also on Reinhart's staff, did some design refinement and then it became our problem. Manufacturing the ornament was again beyond the standard automotive art.

In this case the solution to the problem came from an article Copp had read of a unique manufacturing method used to produce sights for anti-aircraft guns. We investigated the source of the gun sights and their manufacturing method and the procurement problem for the hood ornament was solved, albeit at a cost per ornament which exceeded that of a complete Ford grill.

Elmer Rohn was our heating and air-conditioning engineer and his wind tunnel tests had shown the leading edge of the rear fenders provided the best pressure area for air intake for the air-conditioning system.

Reinhart struggled with this problem for some time and every solution he proposed was rejected by Mr. Ford because they all detracted from the simple beauty of the car. Finally the styling group came up with a simple "nostril" design which Mr. Ford liked. It was simple in appearance but complicated and expensive to build. This beautiful detail would never have been approved on an ordinary vehicle.

The technique used by Reinhart and developed by Stanley Brom in preparing the styling mockup for final approval is an item worthy of mention.

The full-sized mockup was a combination of fiberglass, acrylic sheet, plaster and plated metal parts, completely trimmed out with all details on the interior. There was not a single appearance item that was not accurately presented in the dummy vehicle. The headlights and other lights worked, the front wheels steered and the door locks operated.

Every visual item one would observe if he saw the real thing on a showroom floor was complete. The committee approved the design and it was two years before the car was on the market. During this time not a single styling change was made or even requested.

Over two years later an incident occurred that reminded me of the accuracy of detail on that model. I was in the executive garage and had occasion to move a vehicle parked there. I climbed in and turned the key and nothing happened. Only then did I realize it was the old styling mockup.

Another instance of attention to detail was Harley Copp's insistence that the temperature gauge on the instrument panel read in exact degrees on the Fahrenheit scale and the gasoline gauge read accurately in gallons.

The bumper chrome itself on the Mark II was the subject of some discussion at the specification stage. The specification for chrome plate is written in the number of hours the plated part will withstand a constant bath of salt water. We wanted it to last at least 25 years and the spec was about four times greater than for the then current Lincoln and Cadillac. When the buyer was purchasing chrome plated parts, he told me our specs were impossible, that no vendor could meet them. He said, "My gosh, Gordon, do you realize your specs are four times tougher than Cadillac?" I said, "We are not building a Cadillac. I think you can get the parts if you pay enough for them." Of course, they didn't.

On October 16, 1954, at a meeting of the Lincoln Continental Owners Club, Bill Ford announced there would be a new car as part of the 1955 Ford Motor Company offering. It was to be "a high quality, low volume, prestige automobile to be known as the Continental." Only later was it announced it would be called the Continental Mark II, based on the assumption the earlier cars were termed the Mark I.

At the same time, Mr. Ford also announced the name of Special Products Operations was being changed to the Continental Division. Back at the Trade School, our work was just about wrapped up on the Continental Mark II coupe, which was the only body style in the initial offering. However, our advanced plans called for the addition of at least two models to the line. The latter of these two was a retractable hardtop coupe on which much work was done, but finally appeared not as a Continental, but as the 1957 Ford Skyliner.

Our first scheduled addition to the Continental Mark II line after the hardtop coupe was to have been a formal, chauffer-driven limousine "of quiet good taste." If memory serves me correct, it had a 145-inch wheelbase and an electrically controlled glass partition back of the driver.

One day Harold Ames, who had been my boss at Duesenberg, was in town and called to tell me he wanted his name on the list for a Continental when they were available. I received permission to show him the styling prototype, which at that time was on the blackboard as a full size rendering on black paper. He felt the limousine would be our most important market. His many years as president of Duesenberg gave his comment a high value.

Later, they built a full scale model of the limousine in clay and it was painted black. There was considerable enthusiasm for it and management asked their advertising agency in New York to suggest a name for it.

Front view of the Continental Mark II limousine clay model in full size shows some contemplated changes to the Mark II treatment. A four-door hardtop in modern vernacular, Gordon Buehrig named it the "Berline".

This four-door limousine version of the Continental Mark II, built in 1955, never went beyond this stage, but was to have been the second body style introduced in the Continental Mark II line. Gordon Buehrig gave it its name, the "Berline".

We would often have a short meeting in Bill Ford's conference room on Monday mornings. One day Mr. Ford showed us an expensive folder from the advertising agency with quite a number of names and a lot of prose describing the merits of each name. After we had discussed each name, Mr. Ford said he didn't like any of them. Then he asked each of us to think about it and if we had any suggestions to bring them to the meeting the following Monday. At this second meeting on the subject, I suggested the name **Berline** and everyone repeated it and said it sounded fine. Mr. Ford asked how I came to think of it and I told him it was because that is what it was. The classic name for a personal type

Another rear treatment on the Continental Mark II limousine included a sliding glass panel as later introduced on Mercury, and the squared off spare tire hump.

Rear treatment modifications on the Mark II limousine prototype, with taillights nipped off at the top and a rounded spare tire. This was again the full size clay mockup.

limousine with a division behind the driver and with the front seat trimmed in broadcloth is **Berline.**

I had gone back to my days at Duesenberg and at Cord to make my last contribution to the Mark II project in naming the new version after my earlier Duesenberg Beverly Berline and 1937 Cord Custom Berline. It seemed to fit and had it been built, that is what it would have been called.

There are many more stories which could be told about our search for perfection in engineering the Continental Mark II, but to do so at this time is academic because the car was a commercial failure. Why did this happen? It was explained by the Ford Motor Company to be the lack of a market for a $10,000 automobile. However, it would seem the economics go deeper than that.

A car which had so much going for it had to have something going against it to fail. It is obvious if the Continental Mark II were to be the best car in the world it would have to be built better than any other. It would have to be put together with the same dedication to perfection that was so evident in the design and engineering. In the American labor market it is impossible to find such craftsmen.

Whatever the reason for the demise of the Continental Mark II, it is obvious to all knowledgeable automotive people the Continental Mark III and Mark IV have a strong aesthetic inheritance from the Continental Mark II.

The only convertible Continental Mark II built by the factory was this one which was constructed on a John Reinhart design by Derham Body Company in Rosemont, Pennsylvania, and eventually given to Mrs. William Clay Ford.

The finished product, the Continental Mark II, as introduced for 1956. One of the early cars is shown here in front of Denison University.

14 Final years with retirement . . . the first time

During my last five years with the Ford Motor Company, I had charge of a small group of engineers in the vehicle research Department. We worked on the application of new plastic materials to automotive components. The work involved a good deal of basic research in the environmental requirements imposed on materials when they were used in various areas of a motor car.

We had been on this assignment for several years and had created a good rapport with all the plastic suppliers.

I wanted to do some basic research on plastic bodies. This did not include polyester fiberglass because such bodies had been in production for a number of years by the Corvette Division of Chevrolet, and therefore would not be research.

It was budget time and I submitted a request for an adequate sum for our work for the following year. We were allowed much less than requested and the anticipated research was therefore impossible. I came up with a scheme to stretch our program dollars by working with some of our suppliers' research facilities.

The scheme was this: we would buy a (Ford) Falcon sedan for a test car and underwrite a program of road tests and wind tunnel tests providing the suppliers would agree to supply us with unique plastic body components to replace the stock ones.

Since the entire program was confidential, the names of the companies which participated in the project cannot be disclosed, but they were some of the best in the plastics business.

Our agreement was only verbal but we had complete faith in them and they in us, that the rules would be carried out 100%, and they were.

Each company was assigned one component and they were not required to divulge any more information than they wished about the experimental material or fabrication detail of their part.

We agreed to provide each company with our test data on their component, but only theirs. Since all parts were painted to match the car, the test vehicle looked like a standard production vehicle.

The left door was made of vacuum formed acrylic sheet bonded to fiberglass. The right door was molded of an elastimeric plastic. The engine hood was vacuum formed of expanded ABS (Acrylinitrite Butadine Styrene). The left front fender was fiberglass but in this instance we were testing an experimental painting material which would greatly improve the economics of glass reinforced plastics. The right front fender was stamped from preimpregnated fiberglass sheet in the B stage of polymerization and then cured. The trunk lid was made by an aircraft firm using a unique type of plastic which had great strength along with extreme lightness.

The research cost of each of these parts was underwritten by the participating companies. This gave us a comprehensive program and I was still within my budget.

The test results remain confidential but I can describe some of my work. I drove the car to Scottsdale, Arizona and, using that as my base, I made a number of runs in the triangle from Scottsdale to El Centro, California, to Palm Springs, California, and back to Scottsdale. It was in August and the ambient temperature ranged from 110° to 120°. The car was not air conditioned. Low temperature testing was accomplished in our Dearborn wind tunnel facility at -40°.

The deck lid, as mentioned, had great physical properties but the material and the process made it too expensive to have a future in high production cars. Ford had an active racing program at the time and this sparked my interest. The strength to weight ratio of the material gave it merit in special high speed vehicles. As most enthusiasts know, the "stock car" races are not run with stock cars. There are rules, of course, and one is they must look like stock cars. A body made of this material would only weight about one-sixth that of a stock car metal body, yet would look just like one.

The material was never used in a race car application by Ford, but I do recall on the test car the torsion bar compensator for the steel trunk lid was not changed and when the key lock was turned, the lid would fly up like the snap of a bear trap!

The data we collected was subsequently presented by a Ford engineer to the Society of Automotive Engineers and the paper has since been a standard reference in the industry.

My retirement from Ford was a momentous event celebrated in fitting manner on two separate occasions. The first was a luncheon affair arranged by the group in Research.

I was so excited about retirement that I wanted to do a work of art to commemorate the occasion. We created it that day before lunch .... well, almost.

This part of the event took place in our small shop area in the Scientific Laboratory. Here we had a very expensive piece of test equipment in the form of a hydraulic press capable of exerting a compressive force of 10,000 pounds and of measuring the force in fractions of a pound. It was an impressive machine, fifteen feet high with an electronic instrument panel and control center.

The theme of the party was to smash my alarm clock and this was to be the machine to do it. I brought some champagne and, to set the mood, we had a Swingle Singers recording of Bach.

My Baby Ben alarm clock's original purpose was to get me to work on time at Stutz, thirty-seven years earlier than this occasion. On this day it was running and was set to ring its alarm at 12 noon. It looked like a sacrificial lamb sitting between the two huge platens of the monstrous machine. The champagne corks popped about a quarter to noon the alarm sounded. Herb Estes pressed a button on the machine and the upper platen started downward at an almost imperceptable speed.

The huge round scale measured the pressure in readings up to 1,000 pounds and then from 1,000 to 2,000 by changing the controls. The Baby Ben stopped ringing but refused to die. When the infernal machine exerted the full 10,000 pounds pressure and the clock was still intact, we went to lunch.

The next day John Chaivre and I took the clock to another area where they had a press capable of 40,000 pounds pressure. When we asked the operator if he could crush my alarm clock, he said "how thick to you want it?" I said about a quarter of an inch. Again, the clock defied destruction. Finally, we propped it on edge and the machine was able to create my flattened masterpiece!

Speakers at Gordon Buehrig's retirement party were, from left, Fred Hooven, Fred Woodward, Warren Fitzgerald (at mike), Van Patrick and John Reinhart.

It was shortly after my retirement from Ford in 1965 that I was invited to teach a course in plastics to students at the Art Center College of Design in Los Angeles. Thus, my retirement was short-lived and I was off to Los Angeles and a new career as a college professor.

Gordon Buehrig laughingly displays his masterpiece created on the day of his retirement from Ford on July 23, 1954—his crushed Baby Ben alarm clock.

The Art Center College of Design has trained something over fifty percent of the automobile designers in the world. I taught at the school for five years. My "Chair" was supported by five great companies: duPont, Monsanto, Uniroyal, Union Carbide and PPG Industries.

The Art Center College of Design is one of the finest art schools in the world. It owes its greatness, at least in part, to the fact no one can teach a course there who has not earned his living in the field in which he is teaching.

My course was a very minor part of a student's overall curriculum. It was only a one afternoon a week session for fifteen weeks. This work schedule was ideal for me as it gave me long weekends. Since the course was basically a lecture course, any assignment I gave the students had to be worked out by them on their own time. Considering this, I was gratified at some of the results. The course was considered very successful and a large part of the credit for this was due to the cooperation I had with the sponsoring companies, who sent their most capable people to Los Angeles to lecture to my classes.

I also had great cooperation from the Los Angeles chapters of the Society of Plastic Engineers and the Society of Plastic Industries. These groups helped arrange meaningful plant visitations and provided experts in materials, tooling and manufacturing processes, who visited the school and conferred with students on their projects.

Every new term presented a new group of students and a new challenge. They were productive years.

A California street corner honors two of motoring's immortals.

Appendices

Composite construction | 1

"Composite construction," a term used in Chapter I, means a body with a framework made of hardwood and covered with a metal skin. Hard Maple and Ash were the two woods used for the frames. Oak was often used for floor boards. Most production bodies were covered with steel panels while some high-priced ones and all custom bodies had aluminum panels.

Ford open models and Dodge open and closed bodies were of all-steel construction. It was the concensus of body engineers at the time that quality automobile bodies would always be of composite construction. Evidence of this general belief can be found in the fact Henry Ford and the Fisher Body Company both purchased thousands of acres of timberland so their body production would never suffer from a lack of wood.

All-steel bodies of the time were noisy, difficult to trim, expensive to engineer and required a very expensive tooling program.

Composite bodies, because of their simplicity and low cost of the process, were naturally the choice of many small automobile companies. Conversely, when the industry switched to all-steel bodies, the complexity and high cost associated with the manufacturing method was a large factor in the demise of many of the smaller companies.

When I was at Gotfredson Body, we were developing the tooling, the assembly fixtures and shaper forms for the Wills Sainte Claire sedan. Our engineering information was a full-scale body draft made on special linen-backed drawing paper. It was 20 feet long and six feet wide. It showed all the pieces of the body in orthographic views in their assembled positions on the completed body. Every detail was included, even to the screws to be used in assembly.

Each piece of the body framework had a name and a number. From the engineering information the woodworker produced sample pieces of every part. When this was completed, these parts were assembled on a large surface plate and the result was a complete body frame. Unlike a real body assembly, nothing was glued together and consequently could be disassembled. The sample body frame proved out the engineering and, if any construction improvements were desired, they were easily made.

When the structure was approved by the chief engineer and the parts carefully disassembled, each part was identified by name and number and then coated with shellac. These master parts were then filed in the tool room.

The next job was to build the manufacturing fixtures. There were small fixtures for subassemblies and one large fixture for framing the complete body. The purpose of the assembly fixture was to position the body parts in their exact position relative to their engineering specified location while the wood screws were driven in. Each joint was painted with hot animal glue and the screws secured the body while the adhesive set up. This completed the engineering and tooling for the manufacture of the body frames.

The sample body shop also built the wooden die models for the body panel stampings. At Godfredson, we purchased our sheet metal stampings from another manufacturer.

The manufacturing process was equally simple. Our wood working equipment included several types of saws, jointers, planers, shapers, etc. Each body frame part was scheduled through the mill and was produced as follows.

The superintendent would check out a master part from the tool room and set up his various pieces of equipment to produce it. The run would be for, say, 500 units. The storage and transportation of parts to the assembly location determined the optimum number of parts to be run. The master part was returned to the tool room as soon as the pieces were made and checked for accuracy. The process was then repeated for the next part.

Flanges on the steel panels were drilled with small holes so they could be nailed to the body frames. If a joint between two panels was on the surface of the body, it was covered with a molding. The designer located such details to be part of the overall body design.

The completed body shells were moved to the paint department and from there to the trim department. Trimming a composite body was easy because the hardwood frame made an excellent structure to which the trim could be tacked.

2 Body design terms

While this book is not intended to be a technical treatise on automobile designing, by necessity it does have descriptions of some of the design problems encountered and their solutions.

In telling the story of these details a few terms are used that are standard in the art but may not be familiar to the reader. Listed below are a few of these terms along with their definitions.

ORTHOGRAPHIC DRAWING

Another term for the same thing would be a mechanical drawing. The body draft, which is an orthographic drawing, is, to the body engineer, the same as a set of blueprints of a building are to an architect. The body draft is far more complicated in that an automobile body surface is composed of compound curving surfaces, whereas a building plan describes mostly flat surfaces. Both are orthographic drawings and are similar in that in a single plane the drawing contains all the information to describe a three dimensional object.

BELT

 This refers to a horizontal line at the base of the window area.

BELT MOLDING

This refers to a horizontal molding often located a few inches below the belt line.

GREENHOUSE

This is a common term in body design and it refers to the area on an enclosed car above the belt.

CLASSIC CAR DESIGN

This refers to the general type of automobile architecture in common use before World War II.

CONTEMPORARY CAR DESIGN

This refers to the slab-sided cars that have been produced since World War II.

BODY PILLARS

These are the vertical structural units that provide the door openings. The matching units on the doors are called DOOR PILLARS. Starting at the front of the front door, the pillar is called the A PILLAR; the next one to the rear is called the B PILLAR and the one at the rear of the rear door is called the C PILLAR.

PACKAGE

In automobile design, the PACKAGE is a drawing and/or a set of specifications that determines such things as overall height, width, length, floor height, front and rear seat headroom and many more such points. The PACKAGE, in effect, determines the proportions of the car.

DASH

The panel which separates the passenger compartment from the engine compartment.

INSTRUMENT PANEL

The panel or board on which the instruments which report on engine and automobile performance are mounted. This is often confused with the DASH, as in early days, they were often one in the same.

TOP OR ROOF

The covering over the passenger's head, be it a convertible or closed car.

QUARTER WINDOW

The small window located at the rear of the rear seat side window. If a sedan does not have one, it is known as a BLIND QUARTER.

UPPER QUARTER PANEL

The metal panel which joins the rear of the TOP or ROOF to the main body.

DECK

The panel to the rear of the ROOF or TOP; also the location of the TRUNK LID.

LOWER REAR QUARTER PANEL

This is the name given the rear panel enclosing the back wheels of the Post-War slab side automobile.

FRONT OVERHANG

The name given the distance between the centerline of the front wheel and the very tip of the front bumper.

WHEELBASE

The distance between the centerline of the front wheel and the centerline of the rear wheel.

REAR OVERHANG

The distance between the centerline of the rear wheel and the very tip of the rear bumper.

OVERALL LENGTH

The distance from the tip of the front bumper to the tip of the rear bumper.

TOE BOARD

The angled floor panel on which the brake and clutch pedals and the accelerator are located and on which the driver and front seat passengers rest their feet.

FLOOR

The flat portion of the bottom of the body structure.

AXLE CENTERLINE

The point at the absolute center of the front and rear axle, from which several other dimensions are taken in body construction.

3 Body surface development

In the chapter on designing at Duesenberg, reference was made to Body Drafting. It states the entire body surface of certain Duesenberg bodies was determined by emperical lines I put on small scale body drafts.

The method is not used extensively today because the body contour is now sculptured in a full scale clay model and the body draft is then made to conform to the model. This method would have been economically impossible at Duesenberg. The older method carries over from the era of horse-drawn carriages.

When one man designs an object for another man to build, he must accurately convey his concept with a drawing which the builder can understand. For instance; if I want to have a block of wood made 2" thick, 4" wide and 8" long, I would describe it with a drawing (Figure 1). I'd give it to a carpenter and he would cut the wood to the proper size.

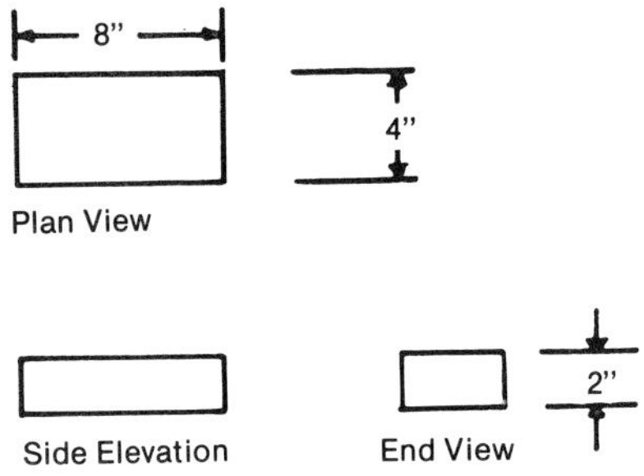

The same principle applies to making an automobile body. One man conceives it and makes a drawing. Another man reads the drawing and builds the body. It is the same thing, just more complicated.

A loaf of French bread is a familiar shape with a compound curved surface. It makes a convenient object to explain body drafting. The overall dimensions can be shown with a simple drawing (Figure 2), but it does not describe the surface contours.

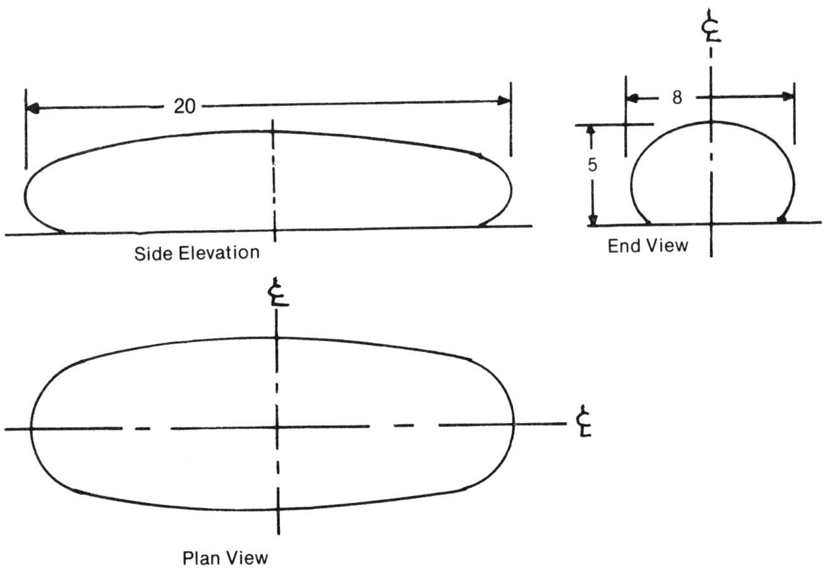

To make a drawing that will describe the surface, visualize the following four steps:

1. Push the loaf of bread through an automatic slicer (Figure 3).

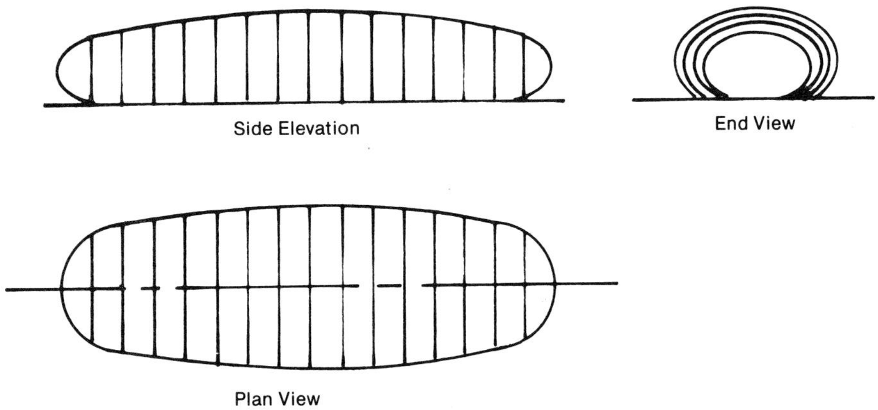

2. Push an identical load through the slicer lengthwise (Figure 4).

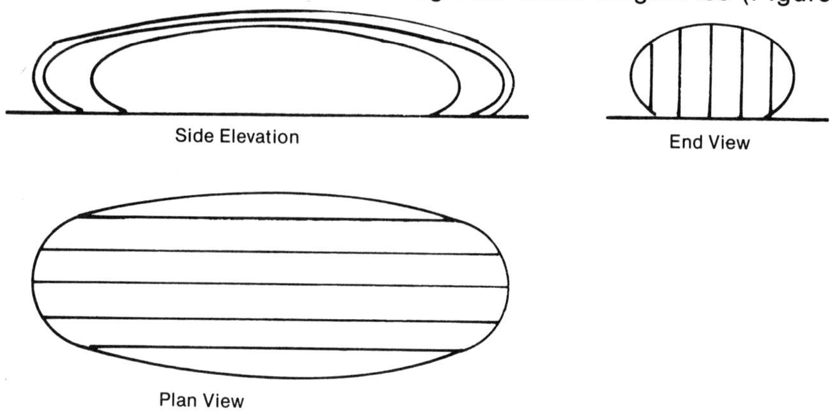

3. Turn a third load up on edge and put it through lengthwise (Figure 5).

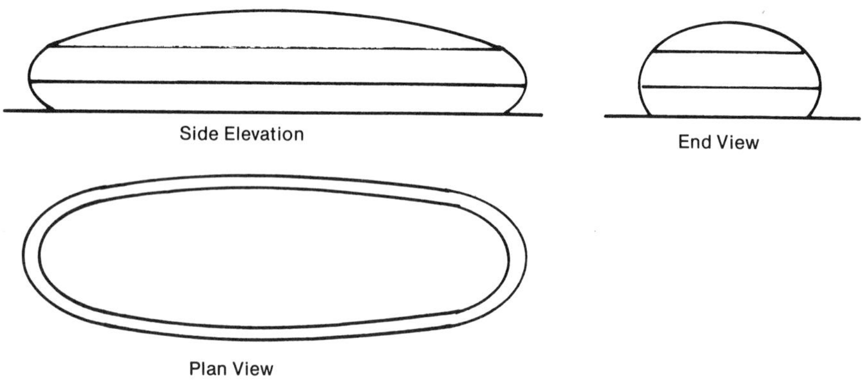

4. Superimpose the three drawings (Figure 6).

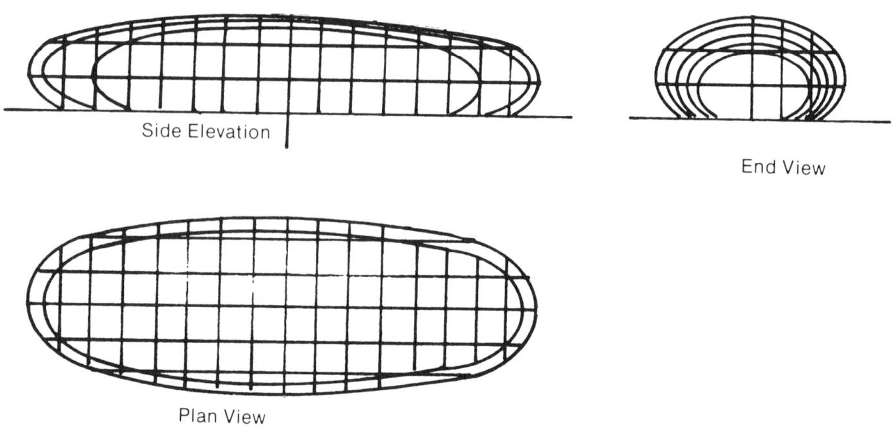

This is a body draft of a loaf of French bread. It describes the surface. Suppose this shape is an object that must be made accurately. I have to make the drawing (Figure 6), so a toolmaker can build a mold. I must do it all on paper because I have no model from which to take templates or cut sections. I have been given Figure 2. All the lines on Figure 6 are determined by the emperical lines on Figure 2 but I have to develop them by body drafting geometry.

The purpose of this exercise is to demonstrate the principal of surface development, which was used in body drafting. It is essential that it be understood before moving on to Three Dimension Engineering, which illustrates new concepts in body engineering. The advantages of a new process are not apparent without knowledge of the old. It is only of academic interest today as body surface development is now largely done on computers.

1. Draw section lines 1 through 8 in the side elevation and plan view. Draw center lines in each view and line "W" in the end view and side elevation. Line "W" is at the extreme width.

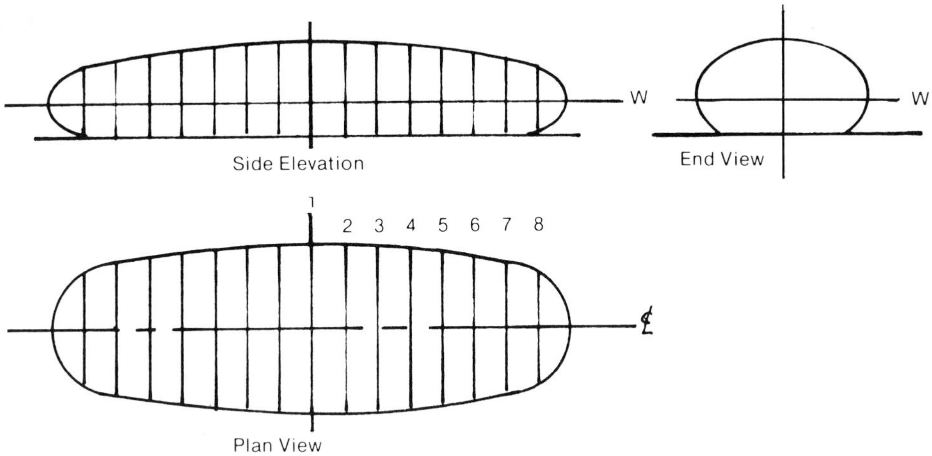

2. Measure the length on line 1 through 8 in Plan View and project these widths on line "W" in the End View. Measure the height dimensions on lines 1 through 8 in the Side Elevation and project these dimensions on the vertical centerline in the End View (Figure 8).

From this exercise only the upper right quadrant of the End View will be developed. The lower right quadrant would be developed by the same method. Since the object is symmetrical about the vertical centerline the left side is a mirror image of the right.

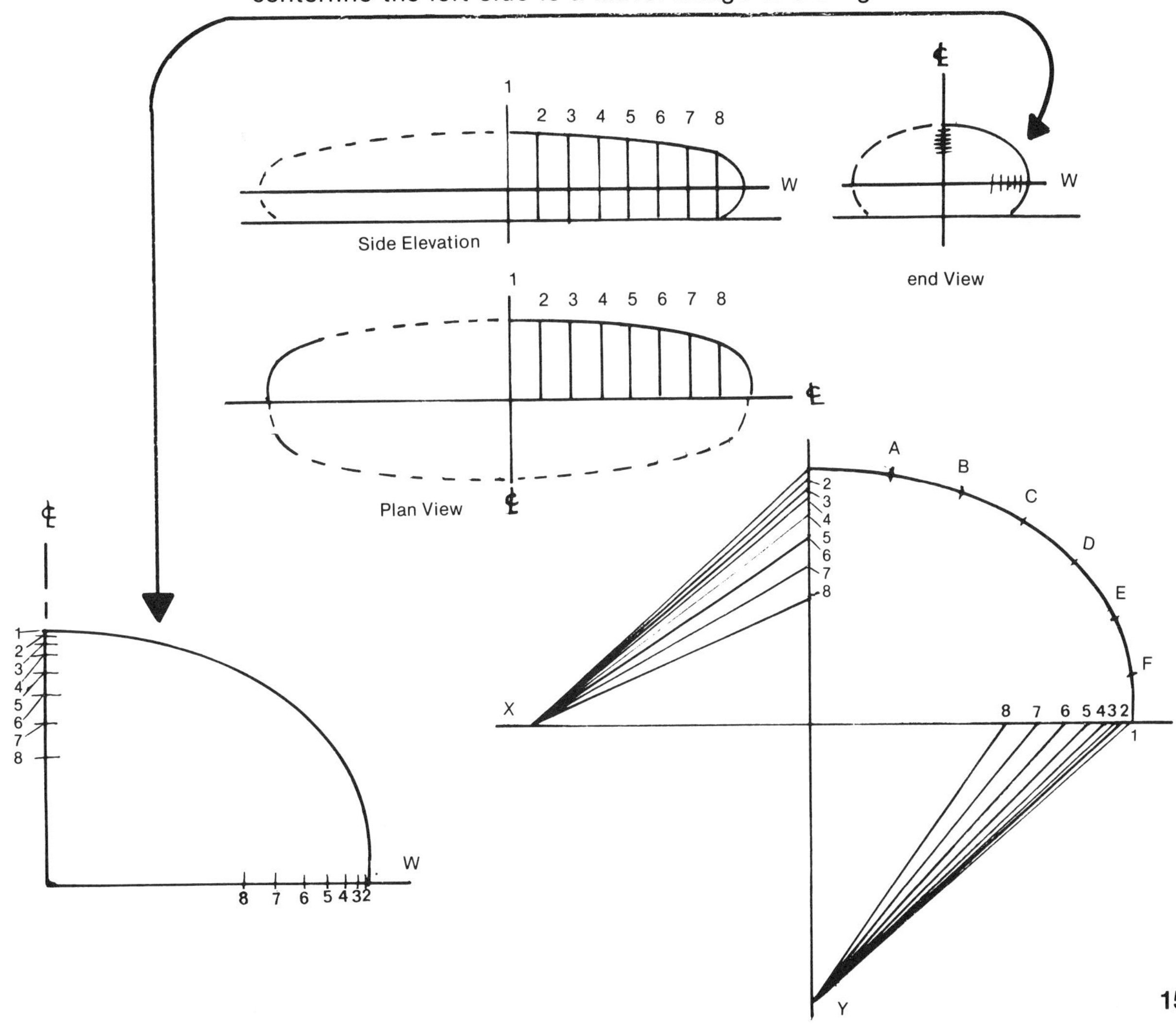

Figure 9 is an enlarged view of the upper right quadrant. Figure 10 shows the vertical centerline extended and the horizontal line "W" extended. The designed shape of the object on the vertical centerline in the Side Elevation (Figure 8) is section 1 in Figure 10. Arbitrary points: A, B, C, D, E and F are located on section 1. The object is to develops sections 2 through 8 that are proportional to section 1.

Figure 11 shows the method. Construct proportional triangles as shown. Points A through F describe section line 1. Similar points, A through F, must be found to describe sections 2 through 8. These sections are also called **Stations.**

Since the method is the same for each, only Station 8 is demonstrated in Figure 11.

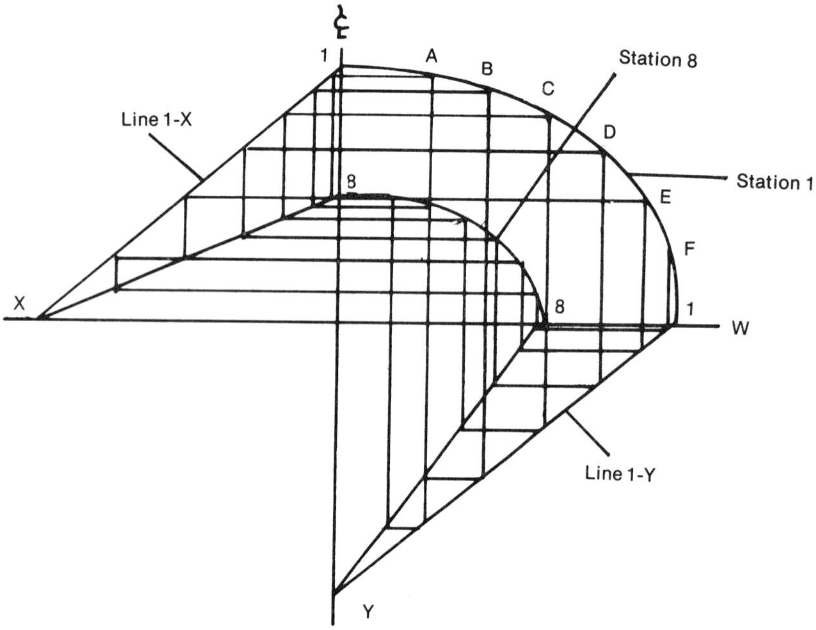

Vertical lines from points A through F are drawn to intersect line 1-Y. Then horizontal lines are drawn from the intersecting points on 1-Y to intersect line 8-Y. From these intersections vertical lines are drawn.

In a similar manner horizontal lines from points A through F are drawn to intersect line 1-X. Then vertical lines to line 8-X. Then horizontal lines are drawn which intersect the developed lines from the other proportional triangle. These intersections are points A through F on Station 8.

Figure 12 shows the developed sections at Stations 2 through 8, above and below line "W".

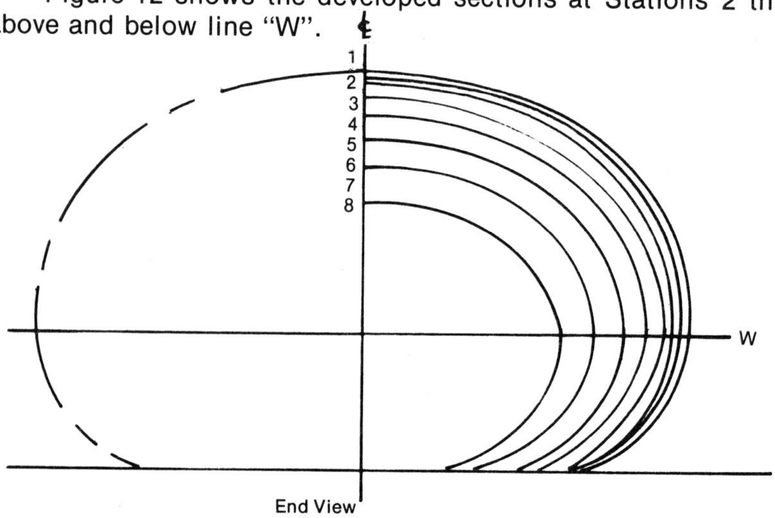

Figure 13 shows the complete draft. Design lines in the Side Elevation are established by projecting intersection points of the contour lines in the End View with the vertical Stations 1 and 2 over to the vertical Station lines in the Side Elevation. Points on the horizontal line 1 in the End View are projected to the Station lines in the Plan View to describe the developed line 1 (above line "W").

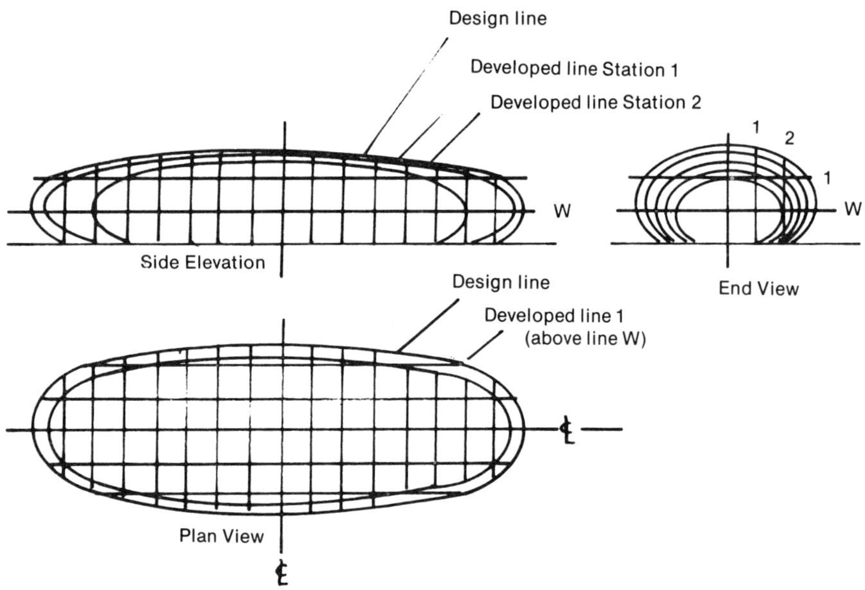

Figure 14 is a perspective sketch of the object showing all the designed and the developed lines.

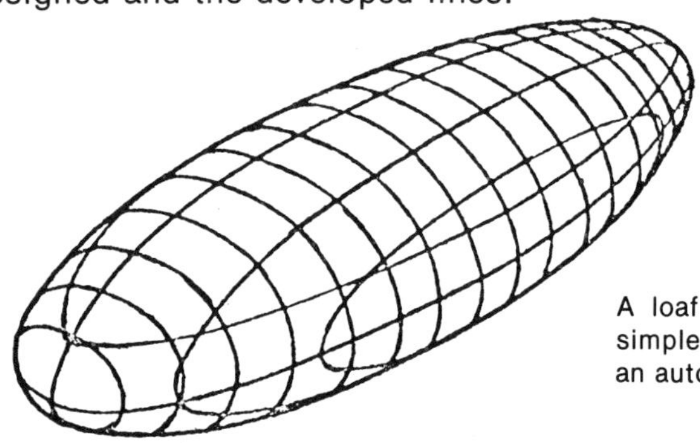

A loaf of French bread is a simple object to compare with an auto body.

Figure 15 is a sketch from ROAD & TRACK showing similar Station lines on a sports car design by Strother MacMinn.

4 | Package evolution: Trend analysis

	Wheel-base	Body Space DASH TO REAR AXLE	Engine Space DASH TO FRONT AXLE	Overall Length
1935 Auburn	127	88.5	38.5	204.7
1936 Cord	125	91.5	33.5	195.2
J - Duesenberg Short Wheelbase	142.5	91.2	51.3	207.5
J - Duesenberg Long Wheelbase	153.5	102.2	51.3	218.5
1972 Luxury 4-Door Sedan (USA)	127	104.8	22.2	225

Earlier in the book, Package, Classic Car Design and Contemporary Car Design were defined as follows:

PACKAGE: In automobile design the Package is a drawing and/or set of specifications that determines such things as overall height, width, length, floor height, front and rear seat head-room and many more such points. The Package determines the proportions of the car.

CLASSIC CAR DESIGN: This refers to the general type of automobile architecture in common use before World War II.

CONTEMPORARY CAR DESIGN: This refers to the slab sided cars that have been produced since World War II.

After the war and the switch to Contemporary auto styling, there came a change in the Package on all American cars. The change was not all bad but it was not all good either. The chart at the top of the page illustrates the change numerically. The 1972 Luxury 4-Door Sedan was chosen as a typical layout of Package dimensions on large American cars. The chart compares these figures with the Auburn, Cord and Duesenbergs in earlier chapters.

Comparing first, the 1972 Sedan with the Long Wheelbase Duesenberg Model - J. The Duesenberg has a 26.5 inch longer wheelbase yet it is 16.5 inches shorter in overall length.

The most frequently used dimension to designate Body Space on a Package is the distance from the Dash (firewall) to the rear axle. Using this dimension to chart shows the 1972 Sedan has 2.6 inches longer body space than the Long Wheelbase Duesenberg, yet its wheelbase is 16.5 inches shorter.

Engine space on a Package drawing is, for comparison purposes, measured by the space between the Dash and the front axle. Comparing the J- Duesenberg again with the 1972 Sedan the chart shows the Duesenberg has 19.1 inches more longitudinal space for the engine yet both cars have engines of comparable cubic inch displacement. The 1972 Sedan is a V-8 and the Duesenberg a straight 8.

The other cars on the chart may be compared in this same manner. From this analysis it is quite apparent the luxury cars of the Thirties were a differently proportioned bunch of automobiles than those being produced in the Seventies.

Is this really progress? I think the answer has to be "yes" if we only consider the most economical use of space and if the American's demand for a tremendous trunk space within the body is satisfied. A rectangular box is a hard shape to beat if economy of space is our goal.

But in automobiles, as in other things, you seldom get anything for nothing. The greatest price paid for the Package of the Seventies is in the lack of character and style. The current crop of automobiles are attractive, they have very roomy interiors and enormous trunk space, but they all look alike! It is difficult to tell a Mercury from a Lincoln or a Chevrolet from a Cadillac.

There are other disadvantages to the current Package layout. Engine compartments are often so crowded that in servicing one component on an engine, removal of several other components to get to it is required, and then they all have to be re-installed. The result is a labor bill greater for all this extra work than for the repair work itself.

Weight distribution suffers by putting an enormous V-8 engine so far forward in the vehicle. This detracts from good ride and handling characteristics and makes the rear brakes less effective. Approximately 80 percent of the braking is done by the front wheels on current cars.

Sheet metal is carried so far forward of the front axle and so far back of the rear axle that even a five mile an hour impact against the bumpers is a costly accident. This was not so on the cars of the Thirties. The slab sided bodies on Contemporary cars are vulnerable to costly accidents on the road and minor ones in parking lots.

Recent television ads by a prominent auto insurance company have demonstrated bumper design was far more effective on cars of the Classic Era than on contemporary automobiles.

On many current designs the widest point on the body is at a point about one-half the distance between the rocker panel and the belt. Above this point the body becomes increasingly narrow and at the top of the door opening the width is less than that of the front seat cushion. When it rains, if the door is open or the window lowered, the outer edge of the cushion get wet. The lower part of the body below the widest point tucks in very rapidly and when the car is driven on wet pavement it is subject to splash from the front wheels. For this reason the lower part of most cars is usually dirty.

With cars of Classic Design the widest point on the body was usually at the belt. Tumblehome was not excessive and the width at the top of the door opening was outboard of the front seat cushion width. (Note the end elevation drawing on the Duesenberg body draft on Page 29.) The narrowest point on the body exterior in this view is at the rocker panel—in those days we called it the sill panel. The running boards protected the body from front wheel splash when the vehicle was driven on wet pavement.

A trend back toward the Classic Car Package might be an improvement over current practice. At least it would be interesting to try it on one line of cars.

Buehrig in miniature | 5

Along with the advent of the old car hobby in the early 1950's in the United States, many sidebar "collection" types of hobbies have grown up with it. This is understandable, as vast collections of the real thing are something reserved for only the very rich.

It is not unusual to find an old car enthusiast who has specialized in the collection of some automotive item, such as literature, radiator ornaments, enamel emblems, photographs ... or model cars.

Model cars are something which has been with us since the advent of the automobile, and some researchers would place the first model car before the first car, claiming models were built by the likes of Daimler and Benz to test their principles before moving on to the more expensive real thing.

The collecting of miniature cars really caught on in the 1950's as, along with the developing old car hobby, the war-developed skill of plastic injection molding was being turned to the model car kit field with great success.

An effort is made here to trace "Buehrig In Miniature" and list all model cars made of Buehrig-designed automobiles, or of automotive design projects he was connected with.

Early models

The first model of a Buehrig design was engineered by Gordon Buehrig himself for the Auburn Automobile Company, and is detailed earlier in Chapter 9 on the design and construction of the 810 Cord. It came late in 1935 when, after being a hit at the automobile shows, Auburn Auto couldn't deliver the 810 Cords which had been ordered, and in many cases, paid for. Harold Ames had Gordon Buehrig do a set of reduced plans for a firm in Buffalo, New York, which manufactured jewelry and was especially good at small bronze castings. A number (not known) of bronze models of the 1936 Cord 810 Sedan were made up in 1/32 scale and were given to customers who had a Cord on order before Christmas 1935. These cost the company approximately $20 each to make and mount on a marble base and today, when one changes hands, it is usually for several hundreds of dollars.

The first model of a Buehrig design came late in 1935 when Buehrig himself drew plans for a 1/32 scale bronze model of the Cord 810 Sedan, which was presented to every customer who had a Cord on order before Christmas 1935. Auburn Rubber Company used the same plans to produce a rubber version of the same car.

A second model of the 1936 Cord 810 Sedan was issued in late 1935 by the Auburn Rubber Company of Auburn, Indiana, who were a well-established toy firm which made toys in approximately 1/32 scale from molded hard rubber. According to Gordon Buehrig, Auburn Rubber Company received a set of the same plans and drawings as sent to the Buffalo, New York bronze casting firm. The rubber version sold for a dime and Buehrig recalls having purchased several gross and sending them to his friends throughout the automobile industry.

Buehrig speculated this may have been the first time an automobile company cooperated in providing a model of their car in the United States.

Another 1/32 scale hard rubber toy car was issued by Auburn Rubber Company in either 1936 or 1937 which, while not an authentic copy, had the outside exhaust pipes of the 1937 Cord 812 and a roof line similar to the very few coupe Cords which were built.

These three examples are the only pre-war models of Buehrig designs known to this writer.

This 1/32 scale hard rubber toy car was introduced by Auburn Rubber Company in 1936 or 1937 and owes much of its parentage to the Cord design. Note outside exhaust pipes.

While the Buehrig-designed Cord 810 was an early replica of a real automobile, earlier versions were made of the Austin Bantam and this, the Pierce Silver Arrow, which was shown at the Chicago Century of Progress Exhibition in 1933.

The post-war models in metal and plastic started to flow in the late 1940's and early 1950's. Buehrig designs, especially the Cord and Duesenberg, were quite popular. These and information known about them are broken down by make and model of car.

Cord

1937 Cord 812 Phaeton - The first postwar Cord model was issued by Revell (USA) in 1/32 scale plastic kit form in 1955. Catalog number is H-73. It was molded in dark blue plastic and is no longer available. 1937 Cord 812 Convertible - Issued in plastic kit form by Pyro Plastics of New Jersey in 1960. Scale was 1/23 and it was molded in red plastic. Catalog number is C-332. It is no longer available.

1937 Cord 812 Convertible - One of the finest Cord plastic kits was issued by Monogram Models, Inc. of Morton Grove, Ill. in 1965. It was molded in white plastic, with interior and engine parts in black plastic and chrome. Scale is 1/24. Original catalog number was PC-130. A new version molded in red plastic was issued in 1972, has catalog number 8204 and is still available.

1937 Cord 812 Convertible - The plastic kit model in large 1/12 scale issued by AMT (USA) in 1965 as part of their Grand Trophy series is the finest Cord model yet issued. It has working doors, lights, steering and hood. It can still be found on the shelves of older hobby shops, having been rather expensive.

1937 Cord 812 Phaeton - In approximately 1970, a firm called Cragstan began introducing inexpensive plastic models in approximately 1/43 scale of some cars being modeled in similar scale by companies such as Lesney of England. These cars, with friction motors, were made in the British Crown Colony of Hong Kong. Among them was a scaled down Cord 812, most likely inspired by the Monogram kit as several pieces are of similar design. It is noteworthy as proportions are fairly accurate. Catalog number is 1140B and it can still be found in novelty stores.

1937 Cord 812 Convertible - The first die cast collector's model of the Cord came from the Italian firm of Dugu in 1971. It was in 1/43 scale and of die cast metal. Catalog number is DU-20 for the top down version, finished in white. A top up version was also offered and came in green with a white top. It is still being made, but is no longer imported.

1937 Cord 812 Phaeton - The latest Cord to be issued came on the market in early 1974 as part of Avon Products, Inc. of New York, N.Y. series of car decanters for their after shave lotion products. It is done in approximately 1/25 scale and the decanter is of tan glass with a black plastic roof. The tail is of tan plastic and can be removed to expose the cap. It is still available.

Buehrig-designed Cord 812 convertibles by, from front to rear, Cragstan, Dugu, Monogram and Avon.

The Italian Dugu model of the Cord 812 Convertible by Gordon Buehrig is done in the collector's scale, 1/43, and is all-metal. It features a removable hood and detailed engine compartment.

The Italian Dugu model of the Cord 812 Convertible was sold in both top up and top down versions. It was not imported for long because of its extremely fragile bumpers.

One of the nicest American kit versions of the Cord 812 Convertible is the Monogram Models, Inc. version in 1/24 scale. It was first issued in 1965 in white plastic, but recently has come on the market again molded in red.

Auburn

As popular as the 1935 and 1936 Auburn 851 and 852 Speedsters are among the old car collectors, it is not surprising this is the only Auburn which has been reproduced in model form. It has not proven as popular a modeler's subject as the Buehrig Cord and Duesenberg designs.

1935 Auburn 851 Boattail Speedster - Pyro Plastics of New Jersey issued the Speedster in 1/23 scale plastic kit form in 1960. It carried catalog number C-331 and was molded in red plastic. Fit of pieces was not exceptionally good. It is no longer available.

1935 Auburn Boattail Speedster - Western Models Ltd. of England, a new firm in 1973, has begun issuing finished 1/43 scale models of rare and unusual cars, made in die cast metal. Their Auburn is catalog number WML-6.

The only model of the 1935 or 1936 Auburn Boattail Speedster by Gordon Buehrig produced in this country was the Pyro plastics 1/23 scale kit which came out in 1960. It was molded in red plastic.

Surprisingly enough, the Gordon Buehrig Boattail efforts haven't been subjected to the modeler's art too often. The only two present examples are, from left, the Pyro Plastics version of the 1935 Auburn 851 Boattail Speedster, and the exotic replica of the 1932 Duesenberg SJ Wittell Speedster by Weymann. This ceramic beauty was issued in special edition by Ezra Brooks whiskey, and was executed by Ronald Hughes, sculptor-in-residence at Chicago's Studio 22.

Duesenberg

The Duesenberg designs of Gordon Buehrig have proven by far the most popular for the model maker's art. In fact, his 1934 Duesenberg SJ Torpedo Phaeton is one of the world's most copied models. It was issued in 1963 initially by Monogram Models, Inc., and since then has been copied by no less than three other model makers.

1929 Duesenberg Model J Phaeton - This is the earliest Model J Duesenberg plastic kit, introduced by Revell (USA) in 1955. It is a 1/32 scale version of what appears to be the Gary Cooper Derham Tourister. Catalog number is H-72. It is no longer available.

1930 Duesenberg Model J Roadster - This plastic kit, in 1/16 scale, was issued in 1962 by ITC (USA) and little else is known about it. Catalog number is 3638. It is no longer available.

1934 Duesenberg SJ Torpedo Phaeton by Weymann - This is the famous Monogram Models, Inc. 1/24 scale plastic kit issued originally in 1963. Main body color was tan, with a dark interior and chrome plated parts. It carried catalog number PC-81 in the early version, but was reissued in 1972 in bright yellow with a black interior, catalog number 8201. Monogram also issued this kit in 1964 for slot track racing. It carried catalog number SR-2406 and had a Tiger X-2000 motor. It is still available.

1931 Duesenberg SJ Phaeton - In 1964 the Hubley Scale Model Division of Lancaster, Pennsylvania introduced their Duesenberg SJ Le Grande Phaeton, which they called a 1931. This seems a bit early for an SJ. Regardless, it is an outstanding model. The assembly kit is made up of all metal parts in 1/18 scale. It carries catalog number 4864 and is still available.

1931 Duesenberg Model SJ Town Car - Hubley (USA) followed one year later in 1965 with a Town Car version of their Duesenberg SJ, again in 1/18 scale and again an all-metal kit. The exact body copied was not identified. It carried catalog number 4868 and is still available.

1934 Duesenberg SJ Sport Phaeton - Renwal (USA) is another company which issued a plastic kit version of the SJ Torpedo Phaeton. Their scale was smaller at 1/48, but they faithfully copied the color scheme of the earlier Monogram version. The kit, catalog number 151, was issued in 1966. It can still be found.

1931 Duesenberg Town Car SJ - Dugu of Italy was the first company to issue a ready made 1/43 scale Duesenberg and they also chose the SJ Town Car, probably inspired by the earlier Hubley kit. It carried catalog number DU-13 and was issued in cream and brown. It became available in 1967. Like the Cord by the same company, it is still made, but not imported.

1931 Duesenberg SJ Dual Cowl Phaeton - A second Dugu 1/43 scale Duesenberg was the dual cowl phaeton in red, also issued in 1967. Catalog number is DU-19 but it is essentially the same car as the town car version. It is still made but not imported.

1931 Duesenberg Formal Sedan - Solido of France, another of the collector car model manufacturers, issued a limousine version in 1967. It carried catalog number 156 and was finished with a black top above the belt line and grey body. Scale is 1/43 for this all metal ready made and it is still available.

1934 Duesenberg SJ Torpedo Phaeton - An inexpensive plastic ready made replica of the SJ Torpedo Phaeton -- another copy of the Monogram kit -- was introduced by a firm called Lucky in approximately 1970. It is about 1/43 scale and done in roughly the color scheme of the Monogram kit -- tan and brown. It was made on the Chinese island of Taiwan and is probable still available in novelty stores. Catalog number is 9005.

1934 Duesenberg SJ Torpedo Phaeton - One of the nicest 1/43 scale ready made Dusenberg collector models is this one built by RIO of Italy. Again, it was most likely derived from the Monogram model, but was nicely done. It is made with top up in blue with a white top, catalog number 45, or in tan with top down, catalog number 46. Both were issued in 1971 and are still available.

1934 Duesenberg SJ Town Car - A most attractive and unusual Duesenberg model appeared in 1973, produced by the WACO Company of Japan. It is a solid state radio, complete with Duesenberg Town Car body, done in approximately 1/18 scale. It is finished in imitation pewter and is still available.

1943 Duesenberg SJ Town Car - Monogram Models, Inc. has introduced a new Duesenberg plastic kit model in 1974, a Town Car, which uses many parts from the earlier Torpedo Phaeton. This one looks like

The Ezra Brooks whiskey bottle version of the 1932 Duesenberg SJ Whittell Speedster is a ceramic design by famous sculpture Ronald Hughes and is done in approximately 1/12 scale. One of the Ezr Brooks Commemorative Collection, the molds were destroyed after the run was completed.

The Buehrig-designed Duesenberg SJ Torpedo Phaeton by Weymann has proven a popular subject for model makers. In this grouping, from left, are examples by Renwal, Lucky (Tiwan), and RIO.

This Duesenberg town car is quite similar to Gordon Buehrig's Brunn Town Car design. It is done in 1/18 scale and is, in fact, a radio.

The two Italian Dugu Duesenberg Model SJ's in dual cowl phaeton and town car configuration may or may not be Buehrig designs. The dual cowl, other than the awkward rear windshield, could have been based on the Derham Tourister.

Buehrig's Brunn Town Car desing, or another unnamed Town Car he designed. It carries catalog number 8027 and is still available in hobby shops.

1932 Duesenberg SJ Speedster - In the late 1960's the Ezra Brooks company, makers of "Real Sippin' Whiskey" introduced a Commemorative Collection of ceramic bottles for their product. One of these was an approximately 1/20 scale replica of the 1932

This replica of the 1931 Duesenberg Formal Sedan is by Solido of France and done in 1/43 scale.

Duesenberg SJ Speedster built for George Whittell by Weymann to Gordon Buehrig's design. This bottle was the work of sculptor Ronald Hughes, who is sculptor-in-residence at Chicago's Studio 22. The bottle was made for Brooks by Heritage China and when the run was completed, the molds were destroyed. It is one of the most difficult of the Buehrig-designed miniatures to acquire.

Studebaker

Gordon Buehrig was in charge of the Raymond Loewy design studio at South Bend, Indiana, which was given credit for the 1947 Studebaker design -- one that carried on for several years. Early postwar Studebaker models are few, as this was also early in the car model game, and only two could be verified.

1946 (?) Studebaker Champion -- This Starlight Coupe version was issued as a metal ready made in 1948 by Gasgay of Belgium. It was done in 1/43 scale and, as can be imagined, is quite rare.

Probably the best known model of a Gordon Buehrig creation is this Monogram Models, Inc. version of the 1934 Duesenberg Model SJ Torpedo Phaeton built by Weymann. This 1/24 scale plastic kit is finished in tan and has served as the starting place for several other models of the same car.

194? Studebaker Coupe - The second early postwar Studebaker was issued by French Dinky Toy. It was issued in the 1948/49 period and was all-metal in approximately 1/43 scale. It had catalog number 24-0.

Continental Mark II

On the Continental Mark II, Gordon Buehrig was chief body engineer. Since it was his last major design project in Detroit prior to his retirement, we include it here.

1956 Lincoln Continental Mark II - The first Mark II kit was issued by Revell (USA) in 1/32 scale plastic kit form in 1956. It carried catalog number H-1209.

1957 Lincoln Continental - In 1957 AMT (USA) issued a Continental Mark II in 1/25 scale plastic kit form. No additional information is available.

1957 Lincoln Continental Mark II Coupe - Mercury of Italy issued an all-metal ready made car in 1959. It was 1/43 scale, painted blue and carried catalog number 4.

1956 Lincoln Continental Mark II - The German Wiking firm makes models more for model railroad layout decoration than anything else, but in 1959 they issued a Continental Mark II Coupe in HO scale. Catalog number is V21 and this is the only Mark II Continental model still available.

1957 Lincoln Continental Mark II - The American ITC firm issued a 1/25 scale plastic kit of the Continental Mark II in 1959. Little is known about this kit.

6 | Whither the Cord

The design was felt to be so outstanding that following the demise of the Auburn Automobile Company in 1938, there had been a scramble to purchase the Cord body, dies, jigs and fixtures -- a scramble that had been won by Norman deVaux, a former Durant executive and builder of the then-defunct deVaux car.

Not once, but twice new life was breathed into Gordon Buehrig's Cord design ... the Hupmobile Skylark and Graham Hollywood came to be.

It is interesting to note in retrospect the basic idea of both Hupp and Graham had been to take Cord's outstanding styling and duplicate it in a mass production automobile. In the final analysis, they failed in total to build as many Hupp Skylarks (354) and Graham Hollywoods (1,859) as Cord built of the original 810's (1,622) and 812's (1,278).

The Skylark and Hollywood were both inexpensive automobiles with production aimed at a mass market. They have, in fact, been termed the poorest products ever turned out by two fine old carmakers. However, they remain sought-after cars today, largely due to their styling. They represent an automotive oddity in that their design was felt so outstanding it survived the bankruptcy of two firms to continue in production.

It would seem the story of the Cord body dies would end here, but it doesn't. One of the most tintillating stories which has been around the old car hobby for years is what ever happened to those valued Cord body does?

Little thought was given them until the old car hobby caught fire in the early 1960's. At that time, speculation started as to what ever happened to them. Early thought was that following cessation of automobile production in September 1940, Graham-Paige had scrapped the dies. Later rumors were the dies had truely been sold as scrap by Graham, but to a Japanese firm, which had stored them through World War II and planned to eventually reintroduce the Cord.

In the Spring 1964 edition of **Automobile Quarterly**, author L. Scott Bailey really started rumors when, in an article on the then-new Cord 8/10 replicar, he flatly stated the original body dies were sold to Japan as scrap, were carefully hidden in Tokyo during the war and "to this day remain reverently crated in the warehouse of the Nissan Company." Cord enthusiasts immediately jumped on this lead, which brought a letter from the Nissan Company stating yes, they had purchased some dies from Graham-Paige, but this had been in 1936 (long before the Graham Hollywood) and those dies had been used for the pre-war Nissan sedan. They said they had searched their records and could find no purchase of Cord body dies and they did not have them.

Motor Trend Magazine was next to add to the mystery when, in May 1966, they reported a new company located in Ft. Wayne, Indiana, had the original Cord body dies and was planning to revive the Graham Hollywood. However, Neill DeMaree, president of the new Graham Company, was quick to state his firm did not have the dies and they were definitely in Japan, possibly with the Honda Company, and a Japanese luxury automobile was being planned using the dies.

In May 1967, the **ACD Newsletter**, official publication of the Auburn-Cord-Dusenberg Club, reported a person who stated he saw the Cord dies stored in a Yokahama, Japan, warehouse in 1960, but had been unable to find out anything about them.

To this date, the final disposition of the Cord body dies remains an unsolved mystery.

Following are examples of the attempts to produce these cars.

The prototype 1939 Hupmobile Junior Six. Management felt this version looked too much like the Cord and it was dropped before introduction.

From the rear, the prototype 1939 Hupmobile Junior Six was almost indistinguishable from the Cord 810. Taillights and a completely new front end were designed before introduction as the Hupmobile Skylark.

The 1939 Hupp Skylark had a 115-inch wheelbase, 101 horsepower six cylinder motor and was billed as a five passenger sedan. It was first shown in the New York and Chicago automobile shows.

The first 1939 Hupp Skylarks come off the prototype assembly line. The man on the far left is Norman deVaux, originator of the project and Hupp general manager.

A factory photograph of the 1940 Hupmobile Skylark four-door sedan. It was built on a 115-inch wheelbase chassis and carried the Hupp six cylinder motor.

A factory publicity shot of the prototype 1940 Hupp Skylark convertible, the only one built.

The prototype 1940 Hupp Skylark convertible. It was the only one built.

The Hupp Skylark advertisement from **Automotive News,** June 10, 1940. Note the phrase "Built to Hupmobile Standards", a reference to the fact the car was built in the Graham-Paige plant.

These illustrations from the 1940 Graham catalog show the Clipper series which was never produced. Only a very few Hollywood convertibles were produced.

The prototype Graham Hollywood looked almost identical to the Hupp Skylark, except for the chromed upper grille treatment.

7 | The replicars

It has been said the greatest compliment which can be paid an artist is to be copied. If such is true, and it would seem to be, then Gordon M. Buehrig must rank as the most highly complimented automotive designer of all time.

The reason is two of his major designs -- the 1935 Auburn 851 Boattail Speedster and 1935 Cord 810 -- have been the subject of what have come to be known as "Replicars" and a factory has recently opened in Mexico which undertakes the complete "remanufacture" of Continental Mark II's, one of Buehrig's final projects at Ford and on which he was chief body engineer.

The remanufacturing project for the Mark II Continental points up an interesting parallel with the Cord replicars, as a project for remanufacturing Auburns and Cords is what ultimately led to the introduction of the replicars.

The first person to attempt this was Glenn Pray who in 1960 purchased the Auburn-Cord-Duesenberg Company of Auburn, Indiana and started to remanufacture Auburn and Cord cars.

Gordon Buehrig is seen here in his basement hobby shop, building a 1/8 scale clay model to check out proportions for Glen Pray's project. At this point in time he inquired of his employer, Ford Motor, if he could help Pray in this way and was told to stop. No help was given after that and the model was never completed.

In the fall of 1962 he produced the Cord Sportsman using the 1936 Cord 810 as his model. As the car was only 4/5 of original size it was dubbed the 8/10. A Chevrolet Corvair transaxle and Corvair six-cylinder engine were used.

It was more than three years before the first production version rolled off the line at the new Broken Arrow, Oklahoma plant on December 17, 1965. The first production 8/10 was presented to Gordon Buehrig.

The Cord Automobile Company floundered, and on August 24, 1966 was placed into receivership. Glenn Pray started work on another of his dream projects, a replica of the 1935 Auburn 851 Speedster; another of Gordon Buehrig's designs.

At the 1964 Sports Car Show at the Henry Ford Museum, this chassis for the Cord 8/10, called a 1964 Cord Prototype, was displayed. This picture shows the installation of the Corvair engine and transaxle unit.

An attempt was made to find a new investor to take over the Cord Automobile Company, but none were found until late 1968 when W. L. Landers formed S.A.M.C.O., Inc. and took over ownership. He quickly turned out 13 more 8/10 Cords using parts on hand. These were powered by the Corvair engine with front wheel drive. At this stage Landers moved operations from Tulsa to New Mannford, Oklahoma. The first two cars here were produced still with front wheel drive, but with a Buick V-6 engine.

Landers then redesigned the 8/10 simplifying some of the expensive steps in production. The pop-up headlights, as fitted to the original Cord, were first replaced with a tunnel type mounting in the fender. Then the molds were modified and the headlight unit was fitted flush in the fender.

The front wheel drive was also dropped in favor of a conventional rear drive and a Ford 289 cubic inch V-8 engine and transmission.

S.A.M.C.O., Inc. was still producing the Cord automobiles in 1970, at that time powered by either a 302 cubic inch Ford V-8 or the 440 Chrysler Magnum V-8 at the owner's option. Price was then pegged at $8,300. The company, which had also begun production of automobile

The 1966 Cord 8/10 convertible as shown at the New York International Automobile Show. The hubcaps had either been stolen or not shipped with the car, as it was displayed without them.

This interior photograph was taken of a new Cord 8/10 at the 1966 New York International Automobile Show. The dashboard reflects Glenn Pray's effort to come as close as possible to the original.

The prototype 1964 Cord 8/10 is seen here along side an original 1935 Cord 810 convertible. This is the first version of the replicar, with Corvair power, front drive, and Royalite body.

air conditioning equipment, claimed to have a 75 dealer-distributor network and to have sold approximately 400 cars.

Meanwhile, Glenn Pray was hard at work on his Auburn Speedster replica project. By as early as September 1966 the A-C-D Club Newsletter had pictures of fibreglas body molds and fixtures and information that Pray hoped to mount the body on a 125-inch wheelbase chassis and fit a 428 cubic inch Ford V-8 engine and driveline.

Pray displayed several of the finished panels at the A-C-D Club's meet in Auburn, Indiana in September 1966. The parts were well finished

Rear view of the prototype 1964 Cord 8/10 convertible. This photograph was distributed to the press by the factory.

Gordon Buehrig was given the first production 8/10 Cord because of his design of the original Cord model 810.

This publicity photograph shows the 1968 S.A.M.C.O., Inc. version of the Cord 8/10, which was shown at the New York Automobile Show that year. The early S.A.M.C.O. versions continued the front drive and used the little Buick V-6 engine, but this was dropped in favor of rear drive and a Ford 289 cubic inch V-8, which is the version shown here. Note also no more hidden headlights.

and appeared to be interchangable with those of the original 851-852 Auburn Speedsters.

On May 20, 1967, Pray "sneak previewed" the new Auburn, dubbed the Model 866, at an A-C-D Club meet in Chicago. The official presentation of the car came on August 2, 1967 at Tulsa, Oklahoma. The car created a sensation.

Pray had turned to Ford components as a basis, probably because Ford had shown more willingness to help the little guy in terms of providing both the hardware and engineering help.

He used the narrowed Ford Galaxie chassis as a foundation and the Ford High Performance 428 Thunderbird engine as a power unit. All running gear is also Ford.

To those who might wonder how Pray was able to fit the original Auburn 851 Speedster body lines on the Ford chassis, he has a ready answer. He reminds one and all that when Gordon Buehrig designed the original 851 Speedster, he was given a set of standard running gear and body components -- the 1934 Auburn chassis and the earlier Auburn V-12 Speedster bodies -- as a starting point. In this respect things worked out well. Pray had an 851 Speedster in his Broken Arrow shop for a complete restoration just about the time the project was getting around to body building. They took molds from all the major body components, revised them as needed to fit the chassis and built up their first body.

To get a new Auburn 866, you fill out an application for purchase, which must carry with it a $2,000.00 deposit against the total purchase price. This was $8,450 when production began in 1968, but is now closer to $12,000. This entitles the purchaser to choose the color of the car and upholstery, which are the only options available.

What does Glenn Pray think of his second Buehrig-designed replicar? "I really like this car. It's more in line with what I feel this kind of personal car should be."

On April 26, 1968, Pray delivered the first production model Auburn 866 Speedster to General E.C.R. Lasher at the Broken Arrow, Oklahoma plant, with much the same ceremony that had heralded the first production Cord 8/10 in December 1965. The 866, incidently, signifies the car has an eight cylinder engine and design work was completed in 1966.

Glenn Pray introduced his Auburn 866 Speedster in fall 1967 and the car was an immediate hit.

This replica Auburn Speedster was shown at the 1973 New York Automobile Show in November. While not easily identifiable, it could be a kit version of the Elegant Motors version of the car.

The Auburn replicar story doesn't end with the Glenn Pray effort. Another firm, Elegant Motors, Inc. of Indianapolis, Indiana, introduced another replica of the 1936 Auburn Speedster in 1973.

With their introduction of what they call an Auburn 856 Speedster, Elegant Motors entered the luxury replicar field.

Their car is built around a modified 1968 Ford V-8 4-door sedan with automatic transmission. A basic assembly kit can be purchased for just over $4,000. The finished version is quite something else. It is available, as of 1974 for $15,995 for the Ford 351 cubic inch V-8 version, and $19,995 with the Corvette 350 cubic inch version. Weight for the Ford version is approximately 3,600 pounds.

Replicars of the Auburn Speedster are also being built in Palm Beach Florida. All three companies are doing well.

Duesenberg remains the one major area of Gordon Buehrig's work where a replicar of one of his designs has not been attempted. The abortive rebirth of Duesenberg attempted by a Duesenberg Corporation headed by Fred Duesenberg in 1965 to 1967 was not a replicar effort. It attempted to create, using designs executed by Virgil Exner, what the Duesenberg might be today had it continued in production. It listed at $19,500 a copy and never went beyond one Chrysler-powered prototype built in Italy by Ghia.

A more serious effort to build a replicar of the Mighty Duesenberg Model SJ was undertaken in 1970 by Bernard Miller of Gardena, California. Using a Dodge chassis and Chrysler engine, he created a replica of the Duesenberg SJ Speedster, a car designed originally by Herbert Newport after Gordon Buehrig had left Duesenberg and gone on to Auburn Auto.

This most attractive replica also has a quality finish and goes for $24,500 a copy. While initial production has been concentrated on the SJ Speedster, owner Miller says he will build custom bodies to order, which leaves the door open for an eventual replica of a Buehrig-designed Duesenberg body.

Rexall
MICHIGAN 59
2255

Accolades | 8

The Cord Corporation never gave Gordon Buehrig credit for his work, with one major exception. The Model 810 Cord was introduced at the New York Show on November 2, 1935. On November 13, 1935 a New York columnist by the name of Lou Schneider published the following item in his syndicated column: "VERSATILE ... The new Auburn Automobile offering is the creation of Count Alexis de Sakhnoffsky. He's the man who designed the streamlined White motor truck. And -- he's the man who styles the buckles on the Pioneer Suspenders."

One of the newspapers which carried the Schneider column was the **Ft. Wayne Sentinel**. When Gordon Buehrig read it he became so incensed he went into Roy Faulkner and demanded this be corrected or he would have his resignation. He took care of it and in a few days this correction was in Schneider's column: "CORRECTION ... Last week your correspondent stated 'The new Auburn Automobile offering is a creation of Count Alexis de Sakhnoffsky.' That isn't so. The model was created by Gordon Miller Buehrig of Auburn, Indiana. Patents for the design are owned by the Cord Corporation."

In the meantime Faulkner had received this telegram, which Gordon Buehrig has in his scrapbook:

"New York N.Y. Nov. 14, 1935
Roy Faulkner, President
Auburn Automobile Company
Deeply chagrined Auburn Item Stop Correction will be made Promptly and fully Please accept thanks calling our attention to error. Richard Waldo President McClure Newspaper Syndicate"

On November 30, 1935, Gordon Buehrig received the following wire from Sakhnoffsky:

"G. M. Buehrig, Director Design Department
Auburn Automobile Company
Re letter: can assure you have never claimed any participation design nineteen thirty six Cord car stop Believe your design was the only refreshing note at the New York Show stop You are free to use this statement in any way you desire.
Alexis de Sakhnoffsky"

There was also an article in the **Ft Wayne Sentinel** which covered the above incident and which concluded with this statement: "The design in question is the one now used on the radically streamlined 1936 Cord."

After this bit of credit Gordon Buehrig's name was never again mentioned in Cord publicity. He was told that to give credit to him would be unfair to all the engineers who worked on the Cord and who deserved just as much credit as he.

Chris Sinsabaugh, then editor of **Automotive Daily News**, did dig out this information from the company and he wrote the following in his column on November 30, 1935.

"Since Roy Faulkner sprung his sensational Cord front-drive at the New York Show it has been gossiped around that the body designing was an out-side job: that is the work had been done by a consultant brought in for the occasion. Now I have it on the authority of Faulkner that the credit belongs to Gordon Buehrig, who has been in charge of designing work at Auburn for two years and who was with Duesenberg several years prior to this. The design is covered by design patents in Buehrig's name, which have been assigned to the Cord Corp."

The New York Museum of Modern Art had a special exhibit in 1951 entitled "Hollow Rolling Sculpture." (The name of the book comes from this and is used with their permission.) Eight cars comprised the show and, in their opinion, it included the best automotive designs of all time. Arthur Drexler, Curator of Architecture at the Museum, wrote Gordon Buehrig about the exhibit. His letter said in part:

"As for the reasons behind our choice of the Cord for exhibition purposes, the catalog I am sending you will, I hope, explain the method of selection. I might repeat here that we all regard the Cord as the outstanding American contribution to autombile design. The originality of the conception and the skill with which its several parts have been realized makes it one of the most powerful designs in the exhibition, and it has, of course, drawn much interest and favorable comment."

Like most engineering personnel involved with the automobile industry, Gordon Buehrig has been a member of the Society of Automotive Engineers since 1930, having been associated with the Detroit Section. As a member of SAE, he also served on their National Council and, in 1954, was named Vice President of Body Activity on the National Council.

In 1958 Jay Doblin, director of the Institute of Design of the Illinois Institute of Technology, started a project in which he hoped to have the world's hundred leading designers, architects and design teachers select what he called the "best designed mass-produced products of modern times." Each designer was to select his top ten choices, and from this Doblin hoped to come up with a list of the 100 best-designed products.

After a series of ballots the final votes were tabulated and the products were listed from 1 to 100 according to the number of votes they received. The list was led off by Olivetti's Lettera 22 portable typewriter and ran through the Gilbert Erector set. This final tabulation was published in an article by **FORTUNE Magazine** in April 1959.

Gordon Buehrig was rather proud to note the 1936 Cord 810, with design credits to him, was placed 14th on this all-time list, sandwiched between Eero Sarrinen's "womb chair" in 13th place, and the Vespa 125cc motor scooter in 15th place.

In 1960, Floyd Clymer Publications of Los Angeles, California, published a book by three French automotive journalists -- Jacques Rousseau, Serge Pozzoli and Maurice Habermacher -- called simply **100 Of The World's Finest Automobiles**. Auburn, Cord and Duesenberg were all three included in the book and with reference to his designs and body styling in the case of Cord and Duesenberg.

Warren Fitzgerald, the late automotive writer, did a biographical sketch in the February 1966 issue of **Road & Track**. The first paragraph reads:

"With his retirement from the Ford Motor Co. in July 1965, Gordon Buehrig became the last of the great individual classic era designers to leave active participation with an automobile company. **Road & Track** has published articles which cite specific phases of his productive career and we now feel it time to chronicle his many contributions to the richness of our automotive heritage. He will best be remembered for his stunning 810/812 series of front-wheel-drive Cords, but he was also responsible for the appearance of more examples of cars loved and collected by classic enthusiasts than any other designer."

He concluded his article with the final paragraph: "Gordon Buehrig's 810 Cord is the only example of the mid-Thirties 'modern design' idiom revered and collected in numbers today. We asked him why he left the comparative security of General Motors to take on what was certainly a questionable project. He replied, 'it was the opportunity to come in and do a complete new automobile -- you can't turn that sort of thing down.' Had he stayed at General Motors through all these years he might be a wealthy man today. But then our world of automobiles would not be quite so rich."

Road & Track has also carried special articles by Warren Fitzgerald on the Cord, Auburn Speedster, Duesenberg Torpedo Phaeton and Duesenberg Whittell Speedster.

In Volume 10, No. 1, 1972 **Automobile Quarterly** there is an article by Howard Darrin called "My American Safari," which is a biographical sketch about his work at the time he was a car designer in Europe. He returned to the United States from Paris in the autumn of 1937. The story goes: " I was rather unfamiliar with the road qualities of the American automobiles. I knew from personal experience that stylists were not entirely to blame for those top heavy cars, and was highly critical of a condition that existed for many years: the dictatorship of Engineering. Many heads of the industry were ex-engineers who dictated policies based on profits-before-logic and tradition-before-progress. But I did find one American car that was a thorough delight to me -- the 1937 Cord 810.

"Gordon Buehrig -- a genius in his own time -- had fostered a masterpiece with that Cord, and he did so, surprisingly, as both the actual creator of the car and as chief of styling -- something nearly unheard of in the U.S. industry. It was always easier for me to follow through on my own ideas because I was never on the payroll of a major manufacturer as a full-time employee. But Buehrig was, and he still managed to see his idea through to production. He put a great amount of energy, ability and hard work into the 810; I ran into some of his models at Connersville in later years -- alternates to what was actually produced -- and all of them were equally outstanding. There is no doubt that his front end treatment made those of the Big Three obsolete, and this was borne out by the 1942 and early postwar Lincoln Continentals, which used a very similar frontal design. As a strong "radiator man," with great dependence on grilles (like Rolls' and Packard's) to enhance my own body designs, I must say I was impressed. I bought a Cord and the car helped me decide to stay and build cars in America."

Latest accolade came September 8, 1973 at the 23rd Annual Old Car Festival at Greenfield Village, Dearborn, Michigan, when Gordon Buehrig received special recognition from the Henry Ford Museum and Greenfield Village "for his distinguished achievements in the field of automotive design." A loving cup commemorating the event was

Gordon Buehrig was honored at the 23rd Annual Old Car Festival at Greenfield Village on September 8, 1973, when the annual event was dedicated to him and Dr. Donald P. Shelley, director of the Henry Ford Museum and Greenfield Village presented him with a cup in recognition of the event. From left, Leslie R. Henry, Curator of Transportation; Henry Ford Museum; Dr. Shelley; and Gordon Buehrig.

presented to him by Dr. Donald P. Shelley, director of the Henry Ford Museum and Greenfield Village.

Cords have often been used in novels and the following are two examples. The first excerpt is from **Live and Let Die** by Ian Fleming, in which we learn something of the automotive tastes of the famous James Bond.

> "The sight of Leiter's car put these thoughts out of mind. Bond liked fast cars and he liked driving them. Most American cars bored him. They lacked personality and the patina of individual craftsmanship the European cars have. They were just 'vehicles,' similar in shape and colour, and even in the tone of their horns. Designed to serve for a year and then be turned in in part exchange for the next year's model. All the fun had been taken out of them with the abolition of a gear-change, with hydraulic assisted steering and spongy suspension. All effort had been smoothed away, and all that close contact with the machine and the road extracts skill and nerve from the European driver. To Bond, American cars were beetle-shaped Dodgems in which you motored along with one hand on the wheel, the radio full on, and the power-operated windows closed to keep out draughts.
>
> "But Leiter had got hold of an old Cord, one of the few American cars with a personality, and it cheered Bond to climb into the low-slung saloon, to hear the solid bite of the gears and the masculine tone of the exhaust. More than fifteen years old, he reflected, yet still one of the most modern-looking cars in the world.
>
> "They swung on to the Causeway and across the wide expanse of unrippled water that separates the twenty miles of narrow island from the broad peninsula sprawling with St. Petersburg and its suburbs."

In another novel, **Passport to Oblivion** by James Leasor, his hero Dr. Jason Love -- a character somewhat similar to James Bond -- expresses his views on the automobile as follows:

> "He finished breakfast and sauntered into the walled garden under the great cedar of Lebanon that gave shade to the north side of the house, to stand looking at his car. Long, low and white, with the March sunshine glittering on its four polished outside exhaust pipes and the chrome mesh ventilators on either side of the hood, the sight of the Cord always gave Love an absurd thrill of pleasure.
>
> "Here was a car built not as a means of transportation but as the fulfillment of a dream. The fact that the venture failed commercially could not detract from the braveness of the attempt or the beauty of the result. Although made in 1937, the Cord contained so many ingenious ideas still accepted as revolutionary that its attraction among enthusiasts remained at a peak. The few Cords left in circulation changed hands at more than they had cost when new: they were of an age, but somehow ageless."

In the movie version of **Passport to Oblivion** the part of Dr. Love was played by David Niven and the Cord in the picture belonged to James Leasor.

The ACD Club

No listing of honors could be complete without mention of The Auburn-Cord-Duesenberg Club, which was founded in 1952 by Harry Denhard -- an Auburn 851 Boattail Speedster owner -- with the expressed restriction on membership: "For those who have never relished the commonplace."

Over the years the club has prospered. Its membership now exceeds 1500. It has acted as "The Fountain of Youth" to the vehicles owned by its members. As time marches on, the members get older but their cars seem to get younger. Many of the vehicles are restored to a quality equal in every respect to their condition when they were new.

Since Gordon Buehrig was the body designer of some of the cars of each make, he has been the recipient of many accolades by the Club. In fact, he was their first Honorary Member.

The Club holds many activities during the year, including three major annual events. One is held in the east, one in California and the third in Auburn, Indiana.

The Auburn Meet is the most spectacular since it involves the town as well as the cars and the Club members. It is always held on the Friday, Saturday and Sunday of Labor Day Weekend.

With the same regularity as the "Swallows come back to Capistrano," the Auburns, Cords and Duesenbergs return to Auburn, Indiana. Of all the car meets in the United States it is probably the greatest of "Automobile Americana." It is the motor car event which compares -- not in size but in spirit -- to such other national events as the Tulip Festival in Holland, Michigan, the Kentucky Derby in Louisville, Kentucky, the Tournament of Roses in Pasadena, California, and the 500 Mile Race at Indianapolis, Indiana.

What greater accolade could a man ask than that his proudest creative achievement be collected and revered during his lifetime!

Credits

Auburn Automobile Co., Auburn, IN.

Automobile Club de l'Ouest, Le Mans, France

Automobile Oldtimers, Inc., New York, NY

Automobile Quarterly, Inc., New York, NY

Bond, John, R., Road & Track, Newport Beach, CA

Bradley, Jim, Detroit Public Library, Detroit, MI

Budd Company, Detroit, MI

DeBeaumont, P.S., N. Stonington, CT

DeLeeuw, Randall, New York, NY

Desert Camera Shop, Palm Springs, CA

Ford, Henry, Museum, Dearborn, MI

Grignon, Photo, Chicago, IL

Harrah's, Reno and Lake Tahoe, NV

Hess & Eisenhardt Co., Cincinnati, OH

International Automobile Photos, Minneapolis, MN

K.A.W. Photos, Cleveland, OH

Kinsman Photos, Buffalo, NY

Langworth, Dick

Lazarnick, N., Commercial Photographers, New York, NY

Manning Bros., Commercial Photographers, Highland Park, MI

Marchese, Daniel V., W. Caldwell, NJ

Multi-Color, Detroit, MI

News Sentinel, Ft. Wayne, IN

Old Cars, Iola, WI

Peckham, John M., Troy, NY

Peoria Journal Star, Peoria, Il

Poole, Ralph, Burbank, CA (jacket photo)

Research Service, Long Island Automotive Museum, Southhampton, NY

Sports Illustrated, New York, NY

Studebaker, South Bend, IN

Time, Inc., New York, NY

Uhle, H.L., New York, NY

Yost, Stanley K, - The Jeffrey I. Godshall Collection, Royal Oak, MI

Index